T0092814

A HISTORY OF BIOLOGY

A History of Biology

MICHEL MORANGE

TRANSLATED BY
TERESA LAVENDER FAGAN &
JOSEPH MUISE

PRINCETON UNIVERSITY PRESS
PRINCETON & OXFORD

Published by Princeton University Press
41 William Street, Princeton, New Jersey 08540
6 Oxford Street, Woodstock, Oxfordshire OX20 1TR

press.princeton.edu

Library of Congress Cataloging-in-Publication Data

Names: Morange, Michel, author.
Title: A history of biology / Michel Morange ; translated by Teresa Lavender Fagan and Joseph Muise.
Other titles: histoire de la biologie. English
Description: Princeton, NJ : Princeton University Press, [2021] | Originally published in French as Une histoire de la biologie by Éditions du Seuil, 2016. | Includes bibliographical references and index.
Identifiers: LCCN 2020035292 (print) | LCCN 2020035293 (ebook) | ISBN 9780691175409 (hardcover) | ISBN 9780691188782 (ebook)
Subjects: LCSH: Biology—History. | Life sciences—History.
Classification: LCC QH305 .M6713 2021 (print) | LCC QH305 (ebook) | DDC 570–dc23
LC record available at https://lccn.loc.gov/2020035292
LC ebook record available at https://lccn.loc.gov/2020035293

British Library Cataloging-in-Publication Data is available

Editorial: Ingrid Gnerlich and Arthur Werneck
Production Editorial: Kathleen Cioffi
Jacket Design: Heather Hansen
Production: Jacqueline Poirier
Publicity: Sara Henning-Stout and Katie Lewis
Copyeditor: Maia Vaswani

This book has been composed in Arno

Printed on acid-free paper. ∞

Printed in the United States of America

10 9 8 7 6 5 4 3 2 1

CONTENTS

PREFACE

MY FIRST REASON for writing a work titled *A History of Biology* is that such a book does not exist, either in English or in French. The only books available are either old,[1] or dated—in the sense that developments in science are viewed through the lens of positivism and its strict adherence to verifiable proof, or they portray the history of biology as resulting from a succession of discoveries building upon one another.[2] Other works focus only on a single topic, such as evolution.[3] Only François Jacob's *The Logic of Life* and more recent works by André Pichot can be excluded from these criticisms.[4] However, these are very personal works and while the choices they make with regard to the themes discussed make for fascinating reading, they are by no means reference works that biologists can consult, to easily obtain the historical perspective they are looking for.

The books that are available are often frustrating in that they exclude entire periods in the history of biology—antiquity for some, while others ignore the twentieth century.[5] Other books are more complete in their coverage of time but exclude whole fields of biological study, such as plant biology or ecology.

1. Nordenskiöld, 1928; Rostand, 1945.

2. Serafini, 2001.

3. Mayr, 1982.

4. Jacob, 1993; Pichot, 1993, 2011.

5. Magner, 2002; Sapp, 2003.

What distinguishes this book is that it is written specifically for biologists and, more broadly, anyone who is interested in contemporary biology—students and those with an interest in science. The reason I wanted to write a book for these readers stems from personal experience and the very real challenge I faced, when giving introductory lectures and courses on the history of biology, in suggesting a book that would provide more in-depth coverage of the topic. Like those courses and lectures, this book seeks to impart a greater understanding of contemporary biology through its history, by explaining how the ideas and models used today were developed, examining the substance and "strengths" and "weaknesses" of these ideas, and avoiding the trap of labeling theories and models as "new," when in fact they have simply been forgotten, only to reappear and gain prominence later. This book seeks to enrich this biological knowledge by situating it within a historical perspective.[6]

A project such as this will surely raise concerns and criticisms. My ambition to cover the entire history of all areas of biology may seem excessive, and, believe me, it is. It goes against the tide of much of contemporary history, which tends to focus on "microhistories," as this level of study allows one to become fully immersed in the interplay of the key actors and to uncover the causes of historical events. The need to summarize history almost inevitably leads to giving major roles to only a handful of figures, whereas scientific advances are truly collective undertakings. Finally, some would argue that a project such as this would too closely resemble the traditional histories of science in which the development of the discipline is explained through a succession of advancements that led to our modern understanding of it.

6. Morange, 2008b.

It is easy to respond to these criticisms. The richness of microhistories does not negate the need for more comprehensive works. Moreover, the past few years have seen more general works spring up, which would have been more difficult to publish 10 or 20 years ago.[7]

This type of history, which leans on the present to look at the past, is not necessarily a return to a "conservative" view that seeks to justify the value of our contemporary knowledge. It is something altogether different: using the past to shed light on the present, not to justify it. The past can help us better understand the present through the parallels that can be drawn between them, but perhaps more importantly through the ways in which they differ. This notion of holding up a mirror to the past works both ways—the present can also illuminate the past. As Marc Bloch pointed out, history is always written from the perspective of the present—based on our current interests and the questions that we are grappling with.[8] Otherwise, why would so many works have been published recently on the history of cultural exchanges and intermixing of communities, if not to explain our current preoccupation with them? In the history of science, as in social history, historians do not seek to provide solutions by highlighting similarities and differences, but by revealing the issues currently being debated.

Let's add that those who wish to reflect on the history of the biological sciences can start only from their own knowledge of science, acquired in junior high or high school, to which information from the media on the current state of science is subsequently added. This is not, however, the case for physics and mathematics, where the complexity of current research and the

7. See, for example, Müller-Wille and Rheinberger (2012).

8. Bloch, 1941–1943.

sophistication of the mathematical tools needed to understand it often lead to teaching these sciences in an outdated way. One's current understanding of the subject is therefore the starting point and point of reference for anyone who is not familiar with the history of biology.

To fulfill our objective, I have chosen to present this history in chronological order. I have broken it up into periods, some of which roughly correspond to centuries. Without fully subscribing to Michel Foucault's view that all scientists are prisoners of the time in which they live and are limited to the views and tools that it provides them, as was reiterated by François Jacob in *The Logic of Life*, I nonetheless believe that the historical context is crucial. Moreover, there is a dynamic to scientific discovery that only a chronological approach allows you to get at.

For each period—antiquity, the Middles Ages, the Arab-Muslim world, the Renaissance, and so on—the chapter devoted to it is divided into three parts. The first, "The Facts," is a description of the changes that occurred in our understanding of biology and medicine during the period in question. The two sections that follow seek to shine a light on the connections that exist between the past and the present. Here, we can break free from the straitjacket of chronology and take a more comparative approach. "Historical Overview" provides a more in-depth look at certain ideas, lines of questioning, and models that were pivotal both to the period being examined and to contemporary biology, and which were briefly covered in the "Facts" section. These themes are selected with an eye to the present, or, more precisely, what they could add to our modern-day understanding, or "episteme," as Michel Foucault was fond of calling it, of biology. Themes are introduced in the context of the period in which the relevant ideas and questions received the most attention. The final section, "Contemporary Relevance," seeks to compare different periods more directly, by pointing out similarities and differences between the past and the present.

This is a history of biological thought, or what is known as a "conceptual history." Developments in techniques and descriptions of experiments and experimental systems have their place, as do social and cultural history, because science is not separate from them. However, providing only the social, economic, and cultural context does not on its own constitute a history of science. Many recent works on the history of science and an even greater number of articles are disappointing for scientifically minded readers in this regard. They describe the social and cultural context in which scientific knowledge develops very well, better than older works, but often have nothing to say about the substance that makes up this knowledge or the theories and models that scientists are working on. The authors of these works sometimes seem to believe that readers are familiar with these topics and don't need to be reminded of them. More often, reading them reveals that the authors themselves do not have a solid grasp of the subject matter and, more worryingly, do not perceive this lack of knowledge to be an impediment to completing the project. This lack of understanding can even be justified or claimed as a badge of honor, as providing the observer with some distance from the subject being studied.

The drawbacks of covering such a long period of time can be compensated for by the advantages of this more comprehensive approach, which allows one to delve more deeply into aspects of knowledge theory that have long occupied the thoughts of philosophers of science and are still being debated to this day. In the case of life sciences, these include the important but also difficult relationships with other disciplines, such as mathematics and physics, and the continuity/discontinuity dialectic in the development of scientific knowledge.

Most historians highlight changes, breaks, and even revolutions in the development of scientific knowledge. Controversies, which represent points where science diverges, are also given much attention. However, despite admirable attempts at examining them, the dynamics of these changes have not always been explored in enough detail. Furthermore, we will see that in the history of science there is continuity; there are some things that remain fixed; and there are explanations, theories, and models that resurface; as well as a longer-term view, analogous to that noted by Fernand Braudel in the economic and social spheres. Why exactly do these common threads exist and what allows them to remain consistent over time? We will have to strike a balance between the historian's tendency to look at a specific point in time while ignoring the longer arc of history, and the philosopher's penchant for highlighting the fixed nature of things, in a way that is often unwarranted.

Some limits have been placed on this book, particularly in order to keep it to a reasonable length. It would be mistaken to think that each period in this book has been given the "best" possible historical coverage, and experts will make easy work of highlighting the limits, omissions, and sometimes even errors herein. They should consider this book a first version, which their critical input will help improve. Certain chapters of this history are more familiar to the author than others, and this bias will certainly not escape readers. However, the contrasting approach of dividing history into periods, and focusing on the study of these distinct periods, introduces artificial boundaries and obscures the continuities that exist.[9] This book does not claim to replace the substantial

9. Le Goff, 2014.

body of scholarly research conducted on the periods covered. It rather opens the door to it, and I provide a bibliography for further reading, including original works, reference works on specific periods or ideas, and articles on contemporary historical research.

I will not discuss the contributions of several civilizations—namely, Indian, Chinese, Meso-American, and South American societies. The first challenge in including these parts of the world is writing about them competently, especially while attempting to keep this history contained within a concise book. However, another challenge has to do with accurately measuring the level of influence that the historical practices and forms of knowledge of these regions have had on the "Western" worldview of the discipline, which is my main focus in this book. Though these civilizations have contributed more to our present biological knowledge than is widely recognized or acknowledged in the Western world, their influence is not, for the most part, a strong presence in the documented history of biology. Consequently, I have considered a wider-ranging and deeper exploration of their influence to be outside of the scope of this particular book—though this is not to say that such an exploration would not be eminently worthwhile.

Another challenge is the close connection between biology and medicine. Progress in physiology and pathology often went hand in hand, so much so that Georges Canguilhem considered pathology as the gateway to the study of physiology and biology. The players involved are often the same: botanists and naturalists have for the most part been physicians, at least until the eighteenth century. Medical training allowed one to discover botany (through the medicinal use of plants) and zoology (through anatomy), and the profession of medicine provided an income, which zoological and botanical work did not.

Nonetheless, there are other medical skills that are not directly relevant to biologists. We will cover only developments in medicine that, in one way or another, shaped the history of biology.

The title of this work, *A History of Biology*, is problematic in itself, as the term "biology" did not appear until the eighteenth century. Had I opted for "A History of the Life Sciences," it would scarcely have been better—according to Michel Foucault, the concept of life as we understand it did not exist until around the same time. Thus, the nature of this project seemingly gives me permission to write unashamedly on the history of biology, starting from modern-day biology, to put together a history that illuminates it. It is therefore not anachronistic to use the term "biology" to describe events that, though not considered "biological" at the time, have contributed in one way or another (sometimes by clashing with previously held views) to modern-day biological knowledge.

I would like to thank all the people who helped me, directly or indirectly, consciously or unintentionally, to write this book. I won't name them—the list would be very (too) long. I would like to thank all of the colleagues with whom I communicated, and all of my students, whose curiosity and pertinent questions have always been incredibly stimulating. I will simply mention Everett Mendelsohn, who was one of the first to know about this project and who strongly encouraged me to pursue it, and my first two proofreaders, Pierre-Louis Blaiseau and Alexandre Peluffo. I would also like to thank Jean-Marc Lévy-Leblond, who persevered in trying to persuade me to write a work such as this for almost thirty years, and everyone at Éditions du Seuil who contributed very effectively to the publication of this book.

Addendum to the First French Edition

I had asked readers of the first French edition to correct errors that would inevitably have crept into this book, and some did, bringing to bear their knowledge and abilities. I would particularly like to thank Jean Vallade, honorary professor of plant biology at the University of Burgundy; Bernard Saugier, emeritus professor of ecology at the Paris-Sud XI University; and Naoki Sato, a biology researcher at the University of Tokyo who translated this work into Japanese.

Addendum to the English Edition

I want to thank Ingrid Gnerlich, who strongly supported the project of translation from its beginning, and Princeton University Press. I also want to thank my translators, Teresa Lavender Fagan and Joseph Muise, who did a difficult but excellent job.

A HISTORY OF BIOLOGY

1

Ancient Greece and Rome

The Facts

The Birth of Biology

It is impossible to pinpoint the precise moment when the first notions of our modern understanding of biology emerged. Our interest in the natural world is not a new phenomenon—a preoccupation with reproduction, birth, and the nature of disease, as well as descriptions of animal and plant species, can be traced back to ancient times. With the establishment of settled communities and the changes brought about by the Agricultural Revolution, an early biological understanding of the world began to develop. Plants were increasingly employed to treat disease, and with their greater use, efforts to describe them progressed, first in China and India and later in the Middle East. The earliest explanations of the formation of the world and of living things originated in the ancient region of Sumer in Asia, and these were taken up by neighboring peoples and reinterpreted in various ways. The practice of divination and, to a greater extent, the embalming of corpses in Egypt helped advance people's understanding of human and animal anatomy.

Overview of Ancient Greek and Roman Biological Sciences

We won't speculate here as to what gave rise to the development of what we call "science" or to attempts to provide rational accounts of natural phenomena in ancient Greece.

In our look at the history of biology in ancient times, the first period, known as the pre-Socratic period, is of little interest to us. Though Pythagoras (580–495 BCE) and Empedocles (490–435 BCE) attempted to provide overarching explanations of the world, their contributions to biology were limited. The influence of outlying Greek colonies that were in contact with Middle Eastern and Indian civilizations was important in these early stages of the development a scientific worldview. In the field of life sciences, two names are worth mentioning: Anaximander (whom we will touch on later) and Alcmaeon of Croton, who, around 500 BCE, carried out dissections and vivisections, described optical nerves and the Eustachian tube, and made the connection between the formation of thoughts and the brain. Conceptual frameworks were developed, which, while not providing a great deal of substance to add to our biological knowledge, would be drawn upon by later authors and shape the way they thought about the world. These included the nature and number of elements and essential qualities, and the notion that souls animated living beings.

Aristotle (384–322 BCE) is without a doubt the father of biology. Indeed, it was not until the second century CE that Galen, a Greek physician working in Rome, would complete and in some cases correct Aristotle's physiological works and the medical works of Hippocrates and his followers, and Aristotle's natural history works would be taken up and distorted by Pliny the Elder in the first century CE. Nonetheless, it is thanks to the latter that these works were passed on and have survived to this day.

Atomists developed their ideas in a parallel fashion, beginning with Leucippus and Democritus in the fifth century BCE, followed by Epicurus in the third century BCE. In the first century BCE, Lucretius would outline the principles of atomism in his poem *De rerum natura* (*On the Nature of Things*), which is the only account of atomism from this period that has survived.[1]

Hippocratic Medicine

Hippocrates (460–370 BCE) and his followers borrowed the concept of the four elements—earth, air, fire, and water—and the four qualities, in opposing pairs—wetness and dryness, and hot and cold.[2] He extended these divisions to the humors, and differentiated blood (produced by the liver), phlegm (produced by the lungs), yellow bile (produced by the gallbladder), and black bile (produced by the spleen). The predominance of one or another of these humors would lead to four different temperaments, and an imbalance would lead to disease.

A unique feature of the Hippocratic medical school of Kos is that it considered nature to be self-medicating, and thus capable of correcting imbalances as they arose. A physician's role was therefore to promote this power in the patient.

In the field of biology, it was the Hippocratic model of embryonic development, which would later be labelled as "epigenetic," that would have the most lasting impact. In this model, the sperm and ova play equal roles in reproduction. These two

1. Lucretius, 1995.
2. Hippocrates, 2012.

kinds of "semen" are formed in various parts of the parents' bodies, and substances produced from similar parts later recognize each other and combine over the course of the development of the embryo, in a process comparable to fermentation. This model allowed for traits acquired over one's lifetime to be transmitted to the next generation.

Aristotle

In *Timaeus*, Plato (428–348 BCE) added little new to the work of his predecessors, associating life with the presence of multiple souls and framing illness as resulting from imbalances. He considered the entire universe to be a living being.

Aristotle's natural history work has always been considered as secondary and subordinate to his work as a philosopher and physicist.[3] More recently, however, historians have reconsidered this view and some have suggested that Aristotle's natural history and physiological work in fact inspired his work as a physicist and philosopher.[4] Aristotle's work in physics and philosophy can be best illustrated in the living world, and without this context it is often difficult to understand.

Aristotle's body of work on natural history is quite substantial. Not only did he put forward one of the first classifications of animals that divided them into species and genera, but his descriptions were also generally very accurate. This precision was drawn from his own observations and experience, but also from conversations with fishers and travelers. This doesn't prevent us from pointing out the more questionable ideas that can be found in his work. He took a particular interest in the behavior of ani-

3. Aristotle, 1955, 1982, 1991.

4. Lennox, 2001.

mals and their lifestyles, and comparisons between human behavior and animal behavior are a recurring theme in Aristotle's writing, in which humans don't always come out on top! However, we generally find support for the notion of a hierarchy of beings in his work, with humans at the top.

Aristotle did write a work on plants, although it has not survived. However, his successor at the head of the school, Theophrastus (371–287 BCE), would do for plants what his predecessor had done for animals. This work was the product of four years' study (between 347 and 343 BCE) undertaken jointly by Aristotle and Theophrastus, which would result in 200 works written by Theophrastus, of which only 2 have survived: *Enquiry into Plants* and *On the Causes of Plants*.[5] Theophrastus separated trees, shrubs, and herbs, and paid a great deal of attention to the environmental conditions that were favorable to plants, which is why he is sometimes considered the father of ecology. However, his descriptions of plants are written very much in relation to human needs. He would, for example, describe the conditions that foster the growth of trees to produce wood that is easy to work. A large part of these works is devoted to medicinal plants and their uses, and points out that it is often through similarities in shape or color that plants reveal their therapeutic uses to us.

In the first century CE, Pedanius Dioscorides (40–90) would complete Theophrastus's work in *De materia medica* (On medical material) by describing more than 600 substances with therapeutic properties obtained from plants.[6] The great renown of this work stems in large part from the fact that it was accessible both in the West and in the Arab-Muslim world.

5. Theophrastus, 1976–1990, 2014.
6. Riddle, 2013.

Aristotle was not particularly innovative in his thinking on the nature of the elements or essential qualities. However, the distinction he made between matter and form was quite an important one in biology. Examples borrowed from biology and medicine will allow us to better illustrate the significance of these ideas and, as we have seen, it was in thinking about living things that these distinctions became apparent. In his view, disease (and death) are rooted in matter, while essence (or what something must be) stems from form. An animal or a plant belongs to a species, and this association is due to its form and not what it is made of, which does not differ from other animals or plants. Similarly, reproduction is seen as a coming together of matter from the female seed and form from the male seed. This union was used to explain how embryonic development was initiated.

The same applies to the distinctions established by Aristotle among four causes—material cause, efficient cause, formal cause, and final cause. The main examples used by Aristotle and his successors to illustrate the different roles of these four causes are borrowed from human activity. In the creation of a statue, the stone or wood represents the material cause, the chisel manipulated by the sculptor is the efficient cause, the formal cause is that which the sculptor wishes to represent (the person), and the final cause is the project of the statue. Similarly, when a physician cures a sick patient by administering plant extracts, the material cause is the extract, the efficient cause is the active ingredient found in the plant, the formal cause is the existence of a state of good health, and the final cause is the physician's desire to cure the sick person.

The notion of the final cause would be vindicated when it was applied to the development of the embryo, while at the same time stirring up more debate. If the formal cause explains why the result of embryonic development will be a cat or a dog, it is the final cause that accounts for the process of embryonic development toward its intended goal—the formation of an adult organism.

These distinctions among the four causes may seem rather counterintuitive to modern readers. Only the efficient cause is still considered a cause. The material cause is no longer a cause, but rather that which causality acts upon. The formal cause is of no particular use and the final cause is incompatible with our nonfinalistic view of the world and particularly the living world, whereby natural processes are not thought to be driven toward some ultimate goal.

Moreover, Aristotle distinguished between three types of soul in living beings—the vegetative soul, which is common to all; the sensitive soul, which is found only in animals; and the rational soul, which is specific to human beings. However, in contrast to Plato, Aristotle believed that souls, and specifically the rational soul, could not be separated from the body.

Unlike his anatomic work, Aristotle's physiology was dependent upon or even "imprisoned" by his philosophical worldview. Thus, due to the prominence he gave to the quality of heat, he believed that the heart, which heated the whole organism, was home to the soul and, for this reason, was the first organ to be formed. For Aristotle, the heat coming from the heart was the work of the soul, and the role of the lungs and the brain was nothing more than cooling.

Aristotle also observed the development of eggs, and, like Hippocrates before him, considered certain steps in this development to be fermentation processes. The quality of his embryologic observations did not preclude him from believing in the spontaneous generation of complex organisms, including certain types of fish.

Aristotle's finalistic views did not, however, go so far as to exclude mechanisms altogether, when, for example, he described the role of tendons in the movement of limbs.[7]

Galen's Physiology

Galen (129–201 CE) was born in Pergamon in modern-day Turkey, where he practiced as physician to the gladiators, and later settled in Rome, where his reputation earned him the title of personal physician to Emperor Marcus Aurelius. Galen's work is characterized by the prominent role he gave to experiments and his strong, sometimes "absolute," finalistic views, whereby natural processes are directed toward some goal.

For Galen, reason and experiments were the two pillars of a physician's work. Galen liked to distinguish his approach from that of more "dogmatic" physicians who denied the importance of experimentation. His role models were Alexandrian physicians from the third century BCE (which we will touch on again later). He practiced animal dissection and, with some restrictions, vivisection. However, in contrast to the Alexandrian physicians and owing to widespread condemnation of the practice, he did not carry out dissections of human cadavers, which would lead to some errors in his anatomic descriptions of the human body. Nonetheless, he made significant contributions to anatomy and physiology, particularly in nerve anatomy and physiology. He demonstrated that the brain was the seat of thought and sight, and situated the soul in the third ventricle (under the cerebellum). He distinguished sensory nerves from motor nerves, and made the connection between spinal cord problems and the sensory and motor deficiencies that result from them.

7. Aristotle, 1991.

The views Galen held on reproduction were a middle ground between those held by Hippocrates and Aristotle on the respective roles played by the man and the woman. Moreover, he was the first to suggest that male and female sexual organs shared a common embryologic origin.

Galen believed that each organ had a specific function and was designed in the best possible way to accomplish it. In his view, organs carried out their functions thanks to the abilities with which they were endowed, to which Galen added many more. As with Aristotle, this finalism did not exclude a more mechanistic approach, and the focus on abilities was sometimes replaced by precise descriptions of the mechanisms involved. Galen fiercely opposed the atomists (more on this later), who believed that organs were not created to perform a function, but rather that it was the nature of the organ that led to its function.

Galen's finalistic worldview, which was linked to his firmly held Stoic beliefs in the existence of a benevolent deity, would allow his work to gain a foothold in a newly Christianized world. Though he liked to think of himself as restoring Hippocrates's work to its rightful place, Galen's work would dominate Western medicine to a greater extent than Hippocrates's until the middle of the nineteenth century.

Pliny the Elder's Natural History

Pliny the Elder (23–79) is known for his tragic death in 79 CE during the eruption of Vesuvius.[8] Wishing to save those in danger but also to learn more about what was happening, he landed with his galley south of Naples and was no doubt asphyxiated by the toxic gases emitted during the eruption.

8. Schmitt, 2013.

Pliny is also famous for having written the 37-volume *Natural History*. His political writings, which were the result of his close relationship with Emperor Vespasian, are much less known, but just as prolific: *Bella Germaniae* (The wars of Germany), which he took part in, and *History of His Times*.

Natural History was the product of knowledge he acquired through book learning and a compilation of prior descriptions, and did not come from study in the field as had been the case with Aristotle. What interested Pliny was not nature itself, but nature that was accessible to and used by humans, and more specifically Roman citizens. When referring to "exotic" animals, he thought it important to mention when the first specimen had been seen by Romans as well as to detail its characteristics. In describing vines, he also detailed methods for preparing wine and their flavor profiles. In keeping with the authors he borrowed from, Pliny endowed animals with human emotions and behaviors: an elephant kneels and prays and studies his lessons, like humans.

Pliny's work is puzzling and can seem to have regressed when compared to Aristotle's, from which he drew much of his inspiration. However, his work had a considerable influence during the Middle Ages and even into the modern era.

The Atomists

The debate around atomism was sparked not by the hypothesis that matter was made up of atoms (indivisible, as the name implies) that were infinite in number but finite in type, as much as the atomists' search for a totally natural explanation of the world, based on chance encounters between atoms. Epicurus (341–270 BCE) built on the ideas developed by Democritus (460–370 BCE) and Leucippus (460–370 BCE), and introduced the notion of *clinamen*—a slight swerve from a straight line in the movement of atoms, which allowed them to preserve their free will. A text by Lucretius (98–55 BCE) is the only work by atomists that has survived.[9] Its poetic form allowed it to endure through the Christianization of society.

Lucretius believed that the primitive Earth was capable of producing all living creatures, including human beings, but also other organisms that have disappeared because they were poorly formed. To survive, specific qualities were needed—speed or visual acuity. When it came to heredity, Lucretius adopted a model that was close to that proposed by Hippocrates—the difference being that he could designate as atoms that which Hippocrates had difficulty naming.

Historical Overview

The Role of Experimentation in Greek Science and Particularly in Life Sciences

You may be asking yourself why we are revisiting this topic. I have already touched on Galen's vivisection experiments, which allowed him to describe different types of nerves and advance our knowledge of the nervous system. The Alexandrian scholars I mentioned before included Erasistratus (310–250 BCE) and Herophilos (310?–250? BCE), who carried out the first quantitative experiments on living things. They weighed them (to estimate the invisible weight lost owing to exhalation), and measured their pulses and how these varied with relation to disease and age. They also conducted dissections of human cadavers, and, according to their rivals, carried out vivisection experiments on human beings.

9. Lucretius, 1995.

However, we must also contrast these achievements with the obstacles that prevented a more systematic implementation of the experimental method—namely, the weight of theoretical reasoning and the priority given to experiments conducted "by analogy." The first of these impediments was felt particularly strongly in medicine, which would very quickly be perceived as a settled discipline, whose principles had been well established since the time of the School of Kos. Even for a thinker such as Aristotle, who was fond of direct observation, it was reason and solely reason from which fundamental principles were derived, which experiments confirmed or occasionally clarified. This preference for reason can be clearly seen in his physiological work: it was not experiments that demonstrated the heart's central role in the organism, but rather reason that allowed us to deduce it thanks to the qualities that this organ possessed.

Experiments also appeared to go against nature. They were a distortion of it, and therefore could not reveal anything about it. It was not only scholars and thinkers in antiquity who held this prejudice—the same criticisms can be leveled against seventeenth-century experimenters. A mistrust of experiments and the hope that reason on its own would suffice to arrive at the correct explanation has probably not been completely eliminated even from the thoughts of modern-day biologists.

The second hurdle was the value accorded to experiments by analogy (or similarity). To illustrate this type of experiment (and explanation), which is particularly common in the Hippocratic corpus, let's look at an example. Why does the female body seem to be more susceptible to water retention, as can be observed in certain diseases? For Hippocratic authors, the answer was simple: because it was less firm. The proof was derived from the following experiment: take raw wool and a sheet of woven wool and place both in the same humid conditions—the raw wool will absorb much more water than the sheet. The result is so obvious that conducting the experiment is often seen as pointless; the experiment itself is a thought experiment.

In *Le chaudron de Médée* (Medea's cauldron), historian Mirko Grmek tried to understand why scientific experiments had not played a major role in Greek science, and particularly in the area of life sciences.[10] Others had advanced the hypothesis that experimentation was curbed by the low status given to technical work (technical trades being reserved for slaves). Grmek came to a different conclusion; namely, that the establishment of an experimental approach is a complex process, which involves several stages to get beyond a groping empiricism. Greek scholars had made it through some of these stages, but not all. There were some attempts at quantification, but it was not widely practiced. What they probably needed most is what Pasteur called an "experimental reflex,"[11] or the widespread recourse to experiments.

Mirko Grmek was right in reminding us that modern-day science and its way of functioning are the result of a long process that was built over several centuries. Greek science was only a chapter in the history of its development.

Anaximander and the Atomists: The Futile Search for Pioneers

Despite repeated warnings from science historians, the hunt for pioneers—the first people to have conducted an experiment or put forward a hypothesis—remains as strong as ever.[12] But this search is of little interest to science historians trying to piece

10. Grmek, 1997.
11. Quoted in Grmek (1997, p. 20).
12. Barthélémy-Madaule, 1979.

together the genesis of an area of scientific knowledge. This is due to the fact that, in most cases, such forerunners were ignored by their contemporaries and successors and thus played no role in the development of the idea. However, more importantly, the notion of a pioneer is a false one, in that it is a retrospective and distorted view that provides the illusion of discerning the beginnings of later ideas in older writings. It is often difficult to disprove the validity of a so-called pioneer. However, the result is always gratifying as it precisely reveals the ways in which our modern-day understanding differs from that of the past.

Ancient Greece still provides fertile ground in the search for these forerunners. The small number of texts (which is why people try to extrapolate things from them) and difficulties translating and interpreting them make it even more so. Let's look at some examples to illustrate the recurring myth of pioneers. Conflicting theories of embryonic development in the seventeenth and eighteenth centuries placed scholars in two camps. Those in the preformation camp believed that organisms were already formed in the egg (or the spermatozoon) and simply grew over the course of embryonic development. Those in the epigenesis camp believed that the organism was formed over the course of its development and did not exist prior to this. Some have claimed that these two models can be traced back to pre-Socratic notions of the universe. For Parmenides (sixth–fifth century BCE) nature was one, and from the beginning contained everything that would later appear. Heraclitus (544?–480 BCE) and others believed that the diversity observed was the result of transformations and that it did not preexist in that which gave rise to it.

One can draw an analogy here, but we have learned to be wary of analogies. The conflict that divided embryologists in the seventeenth and eighteenth centuries was not a revival of an earlier debate, but rather resulted from new observations, particularly in microscopy. While it cannot be denied that this debate can be framed within these older schools of thought, it did not originate from them nor was it shaped by them.

It is with respect to evolution that the search for pioneers has been most actively pursued. When Anaximander (610–546 BCE), in Ionia, described the appearance of life and the formation of the first human beings as fish, did he anticipate our modern-day view of the evolution of the living world?[13] Clearly not, as there are large discrepancies between the scenario he was describing and the account that is widely accepted today. The first discrepancy is the amount of time needed for these processes to run their course. The second and no doubt more important difference is that the transformations described by Anaximander are commonplace in Greek mythology, as indeed they are in the mythologies of various peoples. Developing ideas on the evolution of living forms first required renouncing these fanciful notions. And evolutionary changes would make sense only in the context of our understanding of the stability of living species.

To take the argument further, it is not only the idea of evolution that ancient authors would have had to anticipate, but rather the Darwinian mechanism of evolution. Lucretius described the random recombination of the atoms that generate living beings, leading to misshapen individuals and to others with qualities that allowed them to survive. Lucretius's text does seem modern (or, more precisely, consistent with modern science) in its desire to find a natural explanation for biological phenomena. However, is it truly Darwinian evolution? It seems

13. Kocandrle and Kleisner, 2013.

to me that there are two fundamental differences between Lucretius's view and modern-day thinking. The first is that Lucretius's misshapen individuals disappear—that is to say that natural selection eliminates only individuals that are not viable. This is not in keeping with modern-day thinking on the role of natural selection, even if many have interpreted Darwin's writings in this way, as we will see. The second is that, for Lucretius, individuals that survive do so only because of their particular traits. There is no reference to the central tenet in Darwin's theory that selection acts on relative differences between individuals and not on particular traits. This is a good example of how historical comparisons allow one to refine modern-day thinking.

Other examples of these so-called pioneers must be mentioned briefly, as they have recently found some resonance. The "living universe" described by Plato in *Timaeus* is reminiscent of the living Earth in the Gaia hypothesis that James Lovelock proposed in the 1970s. This is analogous to what we have seen with theories of embryonic development. These are, of course, analogies, and we will see that this idea of a living Earth was a view also held by alchemists. However, to consider the Gaia hypothesis to be simply the revival of an ancient idea does not recognize everything that this hypothesis owes to scientific knowledge accumulated up until 1970.

Similarly, to call Aristotle one of the pioneers of molecular biology by likening the genetic program to the final cause, as proposed by Max Delbrück, one of the fathers of molecular biology, makes little sense. Such a suggestion would not only neglect the novelty of genetic information as an idea, but also be erroneous because for modern-day biologists the genetic program does not represent a final cause but an efficient cause.

Finally, to claim that Theophrastus (or even Empedocles) is the father of ecology is to look at the field in a very simplistic way. As the first farmers no doubt quickly learned, there is much more to ecology than the rather obvious fact that plants don't grow in the same way in different soils, when it is hot or cold, when it is raining or when it is dry.

Could the accomplishments of these alleged pioneers be removed from the scientific record without negatively impacting our comprehension of its history? The answer is less obvious than the preceding remarks may lead one to believe. At least in some cases, these forerunners were able to put together a thought framework within which the models and theories they are credited with having originated could later be understood. I am thinking specifically of the ancient atomists here. Though it would be a stretch to raise their concepts to the status of scientific theory or to claim that the ideas of modern atomists were a continuation of their work, they nonetheless set the stage for new ideas.

Contemporary Relevance

Mechanistic and Molecular Explanations

Models and ways of thinking from antiquity can seem so strange to our modern sensibilities that our first instinct is to dismiss them as irrelevant. However, is this reaction justified? Two types of explanations that still hold sway in biology have their roots in this period: mechanistic explanations and explanations involving the action of ferments.

A mechanistic explanation is an explanation by analogy. That is, the biological phenomenon taking place in the organism is compared to a machine and the explanation hinges on there being mechanisms analogous to those present in machines within the organism. In explanations relating to the action of ferments, it is proposed that phenomena take place that are analogous to those used by humans to transform foods: making bread, alcoholic beverages (wine, beer), cheese, and so on. In a process that is poorly defined, fermentation brings together heat, changes in form and appearance, and small amounts of matter to produce some effect—features that are useful when trying to explain incomprehensible phenomena.

Mechanistic explanations can be found in the writings of Aristotle and Galen. For these authors, such explanations do not account for all physiological phenomena, but they played a role in movement for Aristotle and digestion in the case of Galen, for example. Explanations involving action by ferments come into play in descriptions of embryonic development, but also in explaining the functioning of certain organs, such as the liver or the heart. Both of these types of explanations would have a bright future. Mechanistic explanations would feature prominently in the seventeenth century, without forgetting the action of ferments. The action of enzymes, the successors to ferments, would play a central role in explaining biochemical processes in the first half of the twentieth century. Explanations based around the action of ferments would progressively shift toward molecular ones. Macromolecular mechanisms are now ubiquitous in our explanations of biological processes. The phenomenon of self-organization shares certain characteristics with the action of ferments, including its nearly limitless ability to explain things.

The Role of Analogy

Given the persistent nature of these two types of explanations, we should ask ourselves about the role that analogy plays in the modern-day models we use to explain the natural world. Were we right to scoff at Hippocrates's thought experiments? Analogy is an indispensable tool in science, in particular for the development of models, but it must not, as was the case with Hippocrates, replace experimental facts. It is particularly prevalent in biology, perhaps because it makes use of everyday language. Can we distinguish a good analogy from a bad one? It appears not, as we can know only in hindsight whether the analogy will have advanced our understanding of the phenomena in question.

The disciplines from which analogies are drawn depend on the culture that prevails during the period—i.e., its "episteme," or system of thought and knowledge. This explains why analogies from the past sometimes seem absurd to modern readers. Perhaps in a few centuries some modern-day analogies will appear as ridiculous as those of Hippocrates.

The Beginnings of the Chain of Being

Aristotle was the first to develop the idea of a *scala naturae*, or chain of being, in a scientific way—i.e., that organisms could be more or less positioned along a "ladder" with human beings at the top. This idea would later take root among the naturalists who would follow Aristotle, as well as among embryologists such as Baer, who would characterize embryonic development as a progression from general to specific or from simple to complex.

The concept would not disappear with the rise of Darwin's theory of evolution. The first evolutionary trees naturally positioned human beings at the end of the highest branch. One could argue that it remains influential today, given the position that human beings occupy in many representations of evolutionary trees, or indirectly and in a reactionary way through the often clumsy and unsuitable attempts by those who would like to counter this ancient view and thus make the human line a nearly invisible branch of the evolutionary tree. Their arguments, such as referring to the "small genetic distance" between humans and their closest cousins (chimpanzees), unfortunately often don't make sense from a biological point of view. The chain of being still poisons biological thought.

Pliny's Legacy

In Pliny's writings, it is not uncommon to find distortions of fact or human behavior projected onto animals that he is describing. His works would be nonetheless praised by many naturalists, including Buffon. Do we not have some modern-day Plinys—authors who have poor scientific credibility and who use second-hand information, but who nonetheless receive wide coverage in the media because they know how to frame their ideas for the public to attract attention much better than do scientists, at the risk of sometimes going beyond or even sidestepping scientific knowledge? Regardless of their perceived value, the ideas in Pliny's scientific writing certainly have proved nothing if not remarkably persistent.

Ever-Present Finalism

We should not be too quick to poke fun at Galen's finalism either, which led to justifying the small size of human ears by our need to wear hats! Do we not also indulge in the same finalist thinking when we describe the functions of certain organs? Interpretations of brain imaging are almost as naïve as those put forward by Galen, when they attribute certain cognitive abilities to certain parts of the brain. The same thing occurs when "functions" are attributed to genes and their products. What usually happens is that after a phase of optimistic simplification, genes are found to have multiple functions, which are much more complex than the first observations had led us to believe.

However, finalism had and still has some utility. What Galen proposed was a sort of plan of action—to uncover the functions of different organs—which has proven itself useful in enabling discoveries over the centuries. However, we must nonetheless accept its limitations as demonstrated by experiments.

2

The Middle Ages and Arab-Muslim Science

THE PERIOD STRETCHING from the fifth century to the fifteenth century is generally thought to have been a rather bleak time for the life sciences. Although this period was in fact much more progressive than historians often give it credit, the fall of the Roman Empire and the political and economic chaos that followed meant that a large part of the scientific legacy of antiquity was nevertheless lost. The dominance of religious views of the world would also have a chilling effect on scientific pursuits, relegating attempts to describe and explain natural phenomena to a peripheral activity at best. The discoveries made during this period, be they in the old Roman Empire, Byzantium, or the Arab-Muslim world, amount to only a handful, even if some of the changes ushered in paved the way for future advances, as we will see.

On the flip side, it was also a time of extraordinary cultural exchange. Firstly, there was a movement of knowledge from the Greek and Roman world to the Arab-Muslim world in the eighth century, often with Christians (Nestorians) from Syria acting as intermediaries. From the eleventh century, knowledge flowed in the opposite direction, with the translation of Greek texts from Arabic into Latin, but also works by Muslim thinkers that often drew on these ancient texts. This movement would lead to the golden age of medieval science in the West (the thirteenth century) and eventually to the Renaissance. Modern-day Syria and Iraq were key sites of the first wave of cultural exchanges, and southern Europe, Italy and Spain (Toledo), was the major center of the second.

The Facts

The Arab-Muslim World

The Arab-Muslim expansion brought about political and linguistic unification and the development of trade and bustling urban centers.[1] It was here, in Baghdad, under the Abbasid dynasty in the thirteenth century, that the first attempts to systematically reconstruct the knowledge of the ancients were undertaken. This work was facilitated by the then-recent introduction of paper for writing. The Abbasid rulers wished to put reason back at the heart of teaching, which was no doubt an attempt to take back power from the religious orders. They translated the writings of Aristotle, but also of physicians, including Hippocrates and Galen, as well as texts by other lesser-known authors. When Baghdad lost its dominant position, the movement continued in Egypt and then in Andalusia.

In the fields of medicine and the life sciences, a few notable characters emerged during this period. The first was al-Razi (Rhazes) (865–925), who wrote the first treatise on medicine in Arabic. Avicenna (Ibn Sina, 980–1037), who was born in

1. Rashed, 1997; Djebbar, 2001.

Bukhara, was the author of the well-known *Canon of Medicine*, considered to be the first comprehensive medical treatise. Avicenna's work, like that of al-Razi, is more medical than biological. As in antiquity, dissection was prohibited in the Arab world, more for cultural reasons than religious ones, and thus little progress was made in anatomy and physiology. The absence of representations of the human form in Islamic art precluded the close connection between art and anatomy, as was seen in Europe during the Renaissance. One important exception was the development of the pulmonary circulation hypothesis, which described the flow of blood between the heart and the lungs, by Ibn al-Nafis (1213–1288) at the Mansuri hospital in Cairo, which was the result of reasoning rather than observation.[2] The discovery went unnoticed for the most part, and when Michael Servetus proposed it again in the sixteenth century, it was doubtless an independent rediscovery, and did not draw from the work of Ibn al-Nafis.

Finally, it is worth mentioning Ibn Khaldun (1332–1406) of Kairouan, not for his work on the philosophy of history, but for having anticipated the idea of the evolution of living things. We will not revisit doubts cast on this type of interpretation of his writings here (see chapter 1 for a discussion on this topic).

Arab physicians were also responsible for important progress in surgery. The example most often cited is the "cure" for cataracts, which involved moving the lens with a needle. The technique, which originated in India, was passed on via Greek medicine and was widely used. The success rate was low (estimated at 40%), but must be viewed in the context that the disease often resulted in complete blindness. Other diseases of the eyes, such as trachoma, were well described, and treatments developed.

2. Boustani, 2007.

However, as a rule surgical procedures were not widely practiced, and surgery was mostly limited to simple procedures such as setting fractures. Despite the fact that anesthetic substances such as opium were well known, they were not used in surgery. Postoperative infections took a terrible toll in this period, and the practice of caesarian sections by Arab physicians is just a myth.

More important than these few therapeutic discoveries was the progress made in agriculture and in building hospitals. The description of plants, and particularly medicinal plants, was actively pursued, and more than 400 were added to the list of medicinal plants that had been drawn up in ancient Greece. Botanical gardens were created and agricultural techniques such as irrigation were improved. *Nabataean Agriculture,* a work translated from Syriac at the end of the eighth century, illustrates this rapid change in agricultural practices. The work highlights the selective breeding of, from the animal kingdom, camels, horses, and birds of prey.

The building of hospitals is no doubt the most important contribution of Arab-Muslim civilization, in spite of the fact that their origin (perhaps Persian or Byzantine) remains unclear and the extent that they influenced the later development of Western hospitals remains contested. The first hospital was built in Baghdad between the eighth and ninth centuries, and other hospitals were later built in Damascus, Cairo, Kairouan, and Grenada.

Sites of construction were chosen to promote hygiene and the hospitals were divided by services, including one responsible for people suffering from mental illnesses. They also included a dispensary. Teaching is known to have taken place in these hospitals, but historians cannot agree on the general nature of the instruction or the role it played as a model for others.

Another "modern" aspect of these hospitals is that they were very expensive to run, costs that were thankfully covered through the generosity of princes.

Alchemy is yet another area in which Arab science was particularly active.[3] The alchemy of the Middle Ages and the Renaissance was derived to a much greater extent from Arab alchemy than that of ancient Greece. The topic is somewhat irrelevant to this book, however, despite the fact that a number of scholars, such as al-Razi, Avicenna, and Albertus Magnus, devoted a great deal of time to it. This is because alchemy was concerned chiefly with the transformation of matter at this time; though, as we will see, that would cease to be the case during the Renaissance.

The Middle Ages in the West

There is very little worth noting about the development of the biological and medical sciences in the West before the revival that was ushered in with translation of Arab texts, beginning in the eleventh century and continuing into the twelfth and thirteenth centuries.

This was a period of economic expansion, where forests were cleared and trade and cities developed. Gothic art spread quickly, and with it, more realistic representations of the human body.

The first universities were founded near cathedrals, in Bologna at end of the eleventh century, in Paris and Oxford in the middle of the following century. Medical schools were also established, in Salerno in southern Italy, and later, at the end of the thirteenth century, in Montpellier. The recently rediscovered

3. Joly, 2013.

works of Aristotle were being discussed, and we even begin to see experimental approaches in the field of physics being introduced (or reintroduced).

However, it is very difficult to point to any truly novel ideas or discoveries in the fields of biology and medicine. The vigor that characterized the first half of the thirteenth century soon came up against the tougher line taken by religious authorities and, in the following century, the economic disruption of the Black Death. One novel idea that is worth mentioning, however, because of its later importance, is the description of the nature of hereditary diseases.[4] The term "hereditary disease" first appeared in translations of Arab texts, in particular in Avicenna's *Canon,* though it is impossible to say whether it was an accurate translation of the terms used in the Arabic source text. The idea would spread quickly, and by the thirteenth century the category of hereditary diseases would be well established, featuring alongside epidemic diseases in medical treatises and encyclopedias.

It is also difficult to identify the biological scholars whose work is worth highlighting in this period. The only two who are typically mentioned by historians are Albertus Magnus and Frederick II, Holy Roman Emperor.

Albertus Magnus (1200–1280), a German Dominican, alchemist, and mineral specialist, devoted his 20-volume treatise *De animalibus,* written in 1270, to describing animal species, the reproduction of insects, and the embryonic development of chickens. His work was not terribly original, and his critiques of Aristotle's work were even less so. He was nonetheless one of the first medieval scholars to show an interest in a return to directly observing animals.

4. Van der Lugt and de Miramon, 2008.

Frederick II (1194–1250), who was king of Sicily and then of Germany, was able to combine contributions from different cultures by surrounding himself with a court of scholars. His works included a treatise on falconry, in which he paid particular attention to the birds' habits. In 1241, he issued an edict authorizing the dissection of cadavers in order to advance knowledge of human anatomy. It was said that he supervised a now-famous experiment seeking to determine which of the known languages was the original language of mankind. Children who were not yet old enough to speak were gathered up at his behest and raised without a word being spoken to them, in order to ascertain which language they would instinctively use to communicate. Needless to say, they never spoke a word and the experiment was a failure.

Historical Overview

The example of hereditary diseases is a particularly interesting one in that it represents a double shift—in the use of language and also in ideas surrounding reproduction.[5] It was the first time that the word "heredity," whose meaning up until then had been limited to the transmission of property between generations, was used to denote to the transmission biological traits.

The introduction of the term "hereditary disease" also marks the beginning of a distinction between the creation of a new organism (the term "reproduction" was introduced only in the eighteenth century, by Buffon) and heredity, the transmission of specific traits. This distinction would find its full expression with the birth of the science of heredity at the beginning of the twentieth century, alongside that of embryology.

5. Van der Lugt and de Miramon, 2008.

The spread of these new ideas in the Middle Ages was somewhat limited, however. Hereditary diseases received relatively little coverage compared with other diseases in the writings of the day. Furthermore, the diseases listed as hereditary were not those that we would recognize today, and included gout, leprosy, and melancholy. Neither was a distinction made between hereditary diseases and congenital diseases. This was in keeping with the half-popular, half-scholarly notion that the circumstances of conception played a major role in determining the traits of the resulting child. It was thought that if one of the parents had a particular disease at the time of conception, it could be transmitted to their descendants. This view was also compatible with the notion of the transmission of acquired traits that was widely accepted by adherents of the Hippocratic model of reproduction. The concept of heredity was therefore limited to diseases and was not expanded to include organisms' other traits. We would have to wait until the nineteenth century for the definition of heredity to be broadened to encompass traits beyond diseases.

It nonetheless represented a genuine shift in thinking, and even the confusion between congenital and hereditary diseases was beginning to be questioned in the Middle Ages.

There were a number of forces at play during the Middle Ages that helped to reinforce a material connection between the generations and provided support for the notion of hereditary diseases. These included the Catholic Church's desire to restore Christian marriage to what it deemed to be its rightful place, its insistence that man and woman became one flesh, and the depiction of genealogies in the form of trees (like the Jesse tree showing the lineage of Jesus, which we find depicted on many stained-glass windows). The notion of heredity would also find its way into language and literature around this time, when the terms *prince du sang*, or "prince of the blood," and "royal blood" came into use. In the popular stories of the time, the bastard would become a central character who would reveal his noble origins, which were biological rather than social, by displaying his true qualities and courage.

On the other hand, there were also factors that worked against developing a deeper understanding of heredity. As we have seen, there was very little interest in animals and their heredity, and only a few species were chosen for study, such as birds of prey. Even the notion of race, which itself was based on a more biological view of the human species, would not become important until later. The idea did not take hold among medieval thinkers for largely theological reasons—that mankind was one and that all humans were touched by the original sin and able to seek redemption—and because they remained loyal to Aristotle's ideas. The differences observed in human beings from different geographic areas were thought to be the result of the diverse environments in which they lived. With time, hereditary diseases would progressively be differentiated from congenital diseases, and would continue to grow in prominence until the concept of biological heredity was expanded to include other traits in the nineteenth century. This example shows how the borrowing of concepts from other areas can lead to new ideas. It also goes some way toward diminishing the extent to which the ideas of Renaissance thinkers in this area can be seen as revolutionary. As with other advances, the magnitude of the conceptual leap was not apparent at the time and it represented neither a break with current thinking nor a new phase in it.[6]

Nonetheless, the introduction of the term "hereditary disease" can also be viewed as a return to the etymological origins

6. Le Goff, 2014.

of the word "heritage," which referred to the transmission of land, or, more specifically, the passing on of gardens (*horti* in Latin). It therefore also included the land's ability to produce new growth again and again each year, adding a "biological" dimension to its meaning. Perhaps the medieval translators were aware of this connection when they used the term "hereditary disease."

Contemporary Relevance

Scientific Progress Is Not a Given

The 10 centuries between the fall of the Roman Empire and the Renaissance clearly illustrate that periods favoring scientific development in both the Arab-Muslim world and the West coincided with periods of economic growth and the development of trade and cities—the eighth and ninth centuries in Baghdad and the twelfth and thirteenth centuries in Europe. This connection is further reinforced (in the opposite direction) by the decline of Arab-Muslim science from the thirteenth century onward and of Western science in the fourteenth and fifteenth centuries. In both cases, economic decline was coupled with a rise in religious fundamentalism. In Islamic countries, "Prophetic medicine," based on the teachings of Muhammad, opposed the scientifically based medicine that originated in Greece. Today, we live under the illusion that the position of science in society is incontrovertible and that its status could not be called into question by political or religious powers. However, the history of the twentieth century has shown us that totalitarian regimes, be they Nazi or communist, have slowed or blocked progress in certain areas of scientific endeavor, such as in genetics in the USSR and Eastern Europe and quantum physics in Germany. Indeed, according to a theory put forward by Soviet agronomist Trofim Lysenko (1898–1976), which was supported by Stalin, the environment itself was capable of making hereditary changes to the nature of organisms, even going so far as to change one species into another.[7] This theory led to agricultural disasters and hindered further progress in biology. Likewise, religious ideologies can also call into question established tenets of science, such as the theory of evolution, by finding a sympathetic ear amongst sections of the population and gaining support from leaders.

Less Obvious Contributions to the Development of Science

Though the period described in this chapter was not particularly rich in discoveries, perhaps we underestimate its importance in creating the institutions that would later allow for the development of scientific knowledge: hospitals in the Arab-Muslim world and universities in the West. By the Renaissance, these institutions were already well established and ready to play their role, despite the fact that the pursuit of scientific knowledge had remained almost at a standstill for 10 centuries. Perhaps it is worth recognizing that the criteria by which we judge scientific contributions—those that are used more than ever in selecting candidates for positions and awards—do not represent scientific progress in all its forms. Although discoveries and the development of new theories are important, there are other activities needed for science to function and these also deserve to be recognized. We will look at other examples of these less obvious contributions in this book.

7. Roll-Hansen, 2004.

3

The Renaissance (Sixteenth Century)

THE SIXTEENTH AND SEVENTEENTH centuries are widely considered to be the period during which experimental science as we know it today was born. It was also a period of great social change, which had an impact on society, culture, and particularly art.

In physics, and more specifically in astronomy, there seems to have been an almost perfect continuity between the work of Copernicus, Kepler, and Galileo, and then Newton, in developing a new understanding of the universe. As always, the course of the life sciences ran somewhat less smoothly. The Renaissance can be roughly divided into two periods, with some notable differences. Three major developments in the biological and medical sciences came about in the sixteenth century, which we will include under the umbrella of the Renaissance: notable progress in descriptions of human anatomy, the publication of many illustrated works of zoology and botany, and the coming together of alchemy and medicine. The seventeenth century was characterized by the growing influence of experimentation, mechanistic interpretations of living things, and the introduction of the microscope for observing the biological world. It is sometimes difficult to separate these two centuries, and we must in this chapter and the next sometimes do away with these strict delineations of time.

The Renaissance was a period of strong contrasts. As its name suggest, it was marked by a rediscovery of authors and knowledge from antiquity, and in particular a return to the philosophy of Plato. However, it was also a time when this ancient knowledge was being questioned, and an era characterized by the search for an almost encyclopedic knowledge combining that from the past with new discoveries. The influence of alchemy and alchemists would grow, particularly in medicine, coinciding with a renewed interest in Neoplatonism. A return to literal interpretations of sacred texts would go hand in hand with some progress in anatomy. But despite this progress, there was at least a partial indifference to the physiological and pathological implications of the new discoveries.

The Facts

Progress in Anatomy and Depictions of the Human Body

The work entitled *De humani corporis fabrica* (On the structure of the human body) by Andreas Vesalius (1514–1564), published in 1543, speaks to the progress made in anatomic descriptions during this period.[1] Born in Brussels, Vesalius would make his way to Paris to study anatomy, but clashed with Galen's followers, who dominated the school and were more concerned with studying the master's writings than with precise

1. Canguilhem, 1968; French, 1999; Mandressi, 2003.

anatomic descriptions. He then went to Padua, where he was appointed professor of anatomy at the age of 22.

The illustrations in *De humani corporis fabrica,* engraved by Jan van Calcar, a student of Titian, are what made it such a valuable contribution. They stood out from other works in their precise representations of écorchés (human figures with the skin removed), depicting muscles, blood vessels, and nerves, but even more so for their aesthetic qualities and the ways in which the figures were presented. They were, for example, depicted casually leaning on a chair or looking at a skull, in Renaissance landscapes. Vesalius was not the first to combine anatomy and art. Leonardo da Vinci (1452–1519) had also sketched drawings of his dissections in his notebooks, though these were less impressive.[2] He depicted the skeletal and muscular systems, the eyes, the brain, and the position of the fetus in the uterus. The study of human anatomy and dissection had become required training by this time. Dissections of the human body developed progressively in Salerno, Bologna, Padua, and Montpellier in the fourteenth century. Contrary to popular belief, human dissection was never prohibited by the Roman Catholic Church for religious reasons. It nonetheless was subject to widespread condemnation owing to its perceived lack of respect for the dead, which is why it ceased to be practiced, first in ancient Rome then throughout the Middle Ages, until it gradually reappeared in the fourteenth century. This disapproval had not entirely disappeared, however, and the only "candidates" for dissection were those sentenced to death and foreigners. The small number of cadavers available led to conflicts between anatomists to obtain them.

2. Da Vinci, 2000.

The rarity of cadavers no doubt contributed to making dissections highly coveted spectacles. The fact that the major universities of Europe built amphitheaters that allowed large numbers of spectators to attend these dissections is a testament to their popularity. Vesalius had become a master in the art of public dissection, and had himself perfected the surgical instruments that he used. His dissections drew an audience of more than 500 to the amphitheater in Padua. In other places, dissections were performed by a surgeon, the professor only providing commentary on the results. Vesalius preferred dissecting the body himself and criticized his colleagues for contenting themselves with just reading the texts of the ancients. These dissections drew cries of wonder from the audience, who marveled at the beauty of the human machine, revealing the Creator's admirable work.

There is quite a contrast, however, between this marvelous description and reality. Though dissections were generally done in winter, the bodies did not fare well in the conditions in which they were kept (it took several days to perform a complete dissection). The cadavers would begin to smell quite early in the process and tissues would swell, changing their appearance. Anatomists often had to change cadavers during a dissection, or replace them with animal corpses. Moreover, dissection was a dangerous business—many anatomists injured themselves and died from infections.

Vesalius's 1543 work was unique owing to its aesthetic qualities and presentation, and embodied the widespread interest in human anatomy. Ambroise Paré (1510–1590) also published a book on anatomy—*Anatomie universelle du corps humain* (Universal anatomy of the human body). Vesalius created a prestigious school of anatomy, where he was succeeded by Realdo

Colombo (1510–1559), Gabriele Falloppio (1523–1562), and Hieronymus Fabricius ab Aquapendente (1537–1619), who all made new anatomic observations. Listing all of their discoveries would be tedious for the reader, but it is worth mentioning Colombo's description of the ear and Falloppio's "discovery" of the clitoris and his precise descriptions of the human reproductive system. As we will see, Aquapendente contributed to observations that would lead to the discovery of the circulation of the blood by William Harvey in the following century. These anatomic descriptions also led to advancements in surgical techniques, as exemplified by the work of Ambroise Paré. This included the introduction of artery ligation techniques, replacing the older method of cauterizing the flesh to stop the bleeding when performing amputations. Progress in this area was, however, still hindered by the lack of anesthetics and widespread postoperative infections.

Galen's influence was still strongly felt, despite these breakthroughs and the errors in his work identified by Vesalius and his successors. Vesalius cited Galen 500 times, Hippocrates 300 times, and his own observations only 200 times in his work. He adopted Aristotle's physiology uncritically to interpret his anatomic observations. Perhaps more surprising is the fact that no connection was made or explored between anatomic changes and disease. At best, anatomic study was used to confirm the functions of certain organs or the theory of humors. Curiously, the growing prominence given to mechanistic explanations in the century that followed would not alter this point of view. It was not until the work of Giovanni Battista Morgagni (1682–1771) at the beginning of the eighteenth century that the connection between disease and specific anatomic changes would be studied in a systematic way.

Books on Natural History

Printing allowed images to be reproduced on a large scale from what were initially wood engravings, and led to the rapid publication of many works on natural history.[3] The techniques of printing were first developed in China and later spread to Europe.

One of the forerunners in the publication of natural history books was Conrad Gessner (1516–1565), who lectured in Zurich. His expansive work *Historiae animalium* (History of animals) was produced in four folio volumes, totaling 3500 pages. The book follows Aristotle's classification and the writing reflects Pliny's style, though Gessner could sometimes be critical of the descriptions penned by the ancients. It features illustrations made from woodcuts, including some by Dürer, such as his well-known image of a rhinoceros. Alongside the descriptions, Gessner lists the names of the animals in different languages, the ways in which they are used by humans, cooking methods, medicinal uses, and philosophical and poetic considerations relating to the animal in question. He also included descriptions of mythical creatures amongst those of real animals.

The best known among these authors of illustrated animal histories was Ulisse Aldrovandi (1522–1605), a professor of natural history at Bologna. He created a botanical garden in the city and amassed more than 18,000 pieces in his cabinet of curiosities, most of which he had collected himself during excursions. His works consist of 14 folio volumes, of which only 4 were published during his lifetime, including 3 on birds. The illustrations in his work are even more visually appealing than those by Gessner, though he had a less critical approach than the latter and made more sacrifices in favor of the encyclopedic approach inherited

3. Pinon, 1995.

from Pliny, adding mythological and heraldic notes to the already long list of sections in Gessner's work. Michel Foucault sees Aldrovandi as the model of a sixteenth-century naturalist—which is not meant as a criticism of his work, but rather that his work reflected precisely what was expected of a naturalist at the time.[4]

Although his work as a naturalist is not as extensive as that of Aldrovandi or Gessner, Pierre Belon (1517–1564), unlike the other two authors, would travel to describe animals that were not known in Western Europe. A pharmacist born in Le Mans, France, he traveled through Greece, Turkey, Syria, and Egypt. In addition to his descriptions of plants and their therapeutic uses, Belon published two works on fish—*L'histoire naturelle des étranges poissons marins* (The natural history of strange marine fish) and *La nature et diversité des poissons* (The nature and diversity of fish)—and *Histoire des oiseaux* (History of birds). In his history of birds, he compared human skeletons with those of birds, considered by many to have been the first step toward comparative anatomy, a field that would flourish at the beginning of the nineteenth century. He was the embodiment of a new breed of traveling naturalists who were no longer content to simply describe the animals that others had observed, as was the case with Gessner and Aldrovandi, but rather preferred to be actively involved in discovering and describing new species. Among them, it is worth mentioning Francisco Hernandez (1515–1587), who was born near Toledo in Spain and died in Madrid. After having translated Pliny's *Naturalis historia* (*Natural History*), he spent five years in Mexico. A richly illustrated work by Hernandez on the region's plants and animals would be published posthumously.

Guillaume Rondelet (1507–1566), was a physician from Montpellier and an ardent advocate of dissection, establishing

4. Foucault, 1994.

the city's anatomy amphitheater. He was also a botanist and, like Pierre Belon, would write a history of fish (*Histoire entière des poissons*, "The complete history of fish"). Highly critical of the ancients, he compared organs and their relative positions in different animals.

Though most writers of natural history simply adopted Aristotle's classification, separating animals into oviparous (egg-laying) and viviparous (having embryos that develop inside the body) animals, terrestrial and marine organisms, others preferred to follow the structure found in the book of Genesis, whereby living things are divided according to the order of their creation.

The work of botanists developed in a similar, parallel fashion, and in many ways, the work of Otto Brunfels (1488–1534) did for botany what Aldrovandi's had done for zoology. The parallels extend to the work *De historia stirpium commentarii insignes* (Notable commentaries on the history of plants) of Leonhart Fuchs (1501–1566), which is reminiscent of Gessner's work with its more critical approach. The treatise *De plantis libri* (On plants), written by Andrea Cesalpino (1519–1603), is interesting because of the sometimes-fanciful comparisons that he makes between animal physiology and the physiology of plants, an area of research that would be actively pursued in the century that followed. He also drafted the beginnings of a "natural" classification of plants based on the shapes of their flowers and fruit and the number of seeds.

Also worth mentioning is Flemish botanist Carolus Clusius (1526–1609), who was born in Arras, France, and died in Leiden, in the Netherlands. Clusius published a work on botany and fungi, but also a description of exotic animal species. Another notable character was Felix Platter (1536–1614), a physician in the city of Basel, who was a proponent of a school of anatomic thought based on the work of Vesalius. He also established a herbarium in which he employed novel techniques to improve the preservation of colors in specimens.

Alchemy in Medicine: From Paracelsus to Van Helmont

We have thus far largely avoided the topic of alchemy. Despite it have been practiced in Greece and to an even greater extent in the Arab-Muslim world and the Christian West, it had little impact on the work of naturalists (although many of the scholars previously mentioned, such as Albertus Magnus and al-Razi, were as much alchemists as they were naturalists).

The "Paracelsian revolution" brought alchemy into the practice of medicine and physiology, and after Paracelsus it is no longer possible to ignore the role of alchemy in the history of biology.

Alchemy's long history makes it nearly impossible to define in a way that would encompass all of its practitioners. Certain beliefs were shared by many alchemists, but others less so. Most alchemists considered themselves to be experimenters in their "laboratories," a term that they introduced. They believed that the surest path to knowledge was through experimental work and not through endless commentary on the works of the ancients. One of the goals of the practice of alchemy was the purification of "principles," and the tool most often associated with this was the alembic, a gourd-shaped distilling apparatus.

Most alchemists also shared the belief that there were parallels between the macrocosm (the universe) and the microcosm (the human body). Their interest in astrology derived from this belief. Generally speaking, they believed that similarities in nature were the result of certain signatures that God had placed there for our benefit. For example, medicinal plants would, through their shape or color, carry the signature of their therapeutic uses. Riding the wave of renewed interest in Neoplatonism, alchemists saw the Earth as a living being, whose veins were mineral deposits. These parallels drawn between the Earth and the human body helped to reinforce the manifold connections between the macrocosm and the microcosm.

Alchemists also agreed on the existence of principles—namely, mercury, sulfur, and salt—which represented a break with the four elements of antiquity. These abstract "principles" should not be confused with mercury and sulfur as we know them today.

Alchemy was considered by some of its practitioners to be a hermetic science, intended only for the initiated, while others viewed it simply as a science of matter. The transmutation of metals, and particularly changing base metals into gold, had been a goal of alchemists' work from the start. The philosopher's stone was the tool employed, while also being a means of curing disease and prolonging life. Many alchemists restricted their work to less ambitious projects, however, whose outcomes were more easily attainable.

Paracelsus (1493–1541), who was born in Basel, was one of the "student travelers" typical of the Renaissance, who traveled from city to city in search of new knowledge.[5] Having traveled through France, Germany, Italy, Spain, and Sweden, he was summoned by Erasmus to return to Basel, where he was appointed the city's official physician and began teaching at the university. His colleagues would force his departure a year later, accusing him of having publicly burned the works of Galen and Aristotle to illustrate the need to break with ancient knowledge.

We won't go into detail on Paracelsus's work and ideas on alchemy. The purification of zinc and the introduction of the third principle of alchemy, salt, are often attributed to him. More important was his role in establishing closer ties between alchemy and

5. Paracelsus, 2008.

medicine. He applied alchemy's empirical traditions to his work and praised ideas and recipes from folk medicine. As a result of this work, he would develop laudanum, a drug made from opium that was widely used in medicine until the twentieth century. Though he still believed that diseases resulted from imbalances, the types of imbalances differed from those put forth by the ancients. For him, illness resulted from an imbalance of the principles and, as such, a physician would treat the patient in much the same way as an alchemist would, to purify the organism. The importance of metals in the practice of alchemy and the connections made between the macrocosm and the microcosm led to metals being used as and turned into medications—with some success, as exemplified by antimony. In this way, Paracelsus broke with a 1000-year-old tradition (except in China, where it never took hold) of sourcing medications only from the living world.

Paracelsus devoted a great deal of time to the study of poisons, which he argued were responsible for aging. He posited that it was the dose and not the nature of the substance itself that made it a poison, and thought that diseases were generally caused by external agents, which he was able to confirm through his work on diseases in miners. He was also the first to touch on the role played by the unconscious in the development of disease.

For all of these reasons, Paracelsus is considered to be the founder of chemical pharmacology, toxicology, and even psychoanalysis. Although his radically new approaches are difficult to square with his obscure texts, which are full of metaphysical and religious contemplation.

Almost a century later, Jan Baptista van Helmont (1579–1644) would build on Paracelsus's work, while also at times deviating from it somewhat.[6] He made water the fundamental principle

6. Pagel, 1982.

from which all of the other principles originated—a view he based on his belief in alkahest, a solvent capable of dissolving any other substance. He carried out a now well-known experiment on a willow tree, in which he demonstrated that the tree would grow and increase in weight by simply watering it with pure water. This example illustrates our reluctance to appreciate the value of experiments carried out by alchemists. Even though the conclusions he eventually reached were incorrect, the experiment was well conducted and its results were quantifiable. Indeed, the experiment may have in fact sown the seeds of the principle of mass conservation that would later be explained by Lavoisier and Mikhail Lomonosov (1711–1765).

We run into the same difficulties assessing Van Helmont's work. He was the first to describe carbon dioxide (which he referred to as *gas sylvestre*, or forest gas). In fact, it was Van Helmont who first coined the term "gas" in the scientific literature. Long before Claude Bernard's work, he distinguished the roles of acid and fermentation in the stomach's digestive process. Like Paracelsus before him, he was interested in the ways in which our thoughts affect our bodies, anticipating the development of psychosomatic medicine.

He clearly stated his rejection of scholasticism's insistence on traditional doctrine, and insisted on the need to return to experimentation and observation of nature. The importance he accorded to odors and distinguishing among them is indicative of his familiarity with laboratory work. However, his writings are nearly as obscure as Paracelsus's, and just as laden with references to metaphysics. He believed that fermentations were active owing to their spiritual principles, which he called *archeus*, and that they were therefore the result of seeds that God had scattered in nature. He posited that knowledge, though supported by experiments, was achieved through flashes of spiritual inspiration, and thus remained inaccessible to heretics.

Perhaps we should also include Girolamo Fracastoro (1478–1553) among the alchemist-physicians. He published *De contagione et contagiosis morbis* (On contagion and contagious diseases) in 1546, in which he put forward the idea that contagion is due to the transmission of seeds. He is seen by many as a forerunner in the development of microbial theories on the origin of disease.[7] Fracastoro deserves credit for having come up with a general explanation for the phenomenon of contagion, between plants as well as between animals and human beings, and for having noted its specificity—that is, a disease spreads only within a single species, with only one or a few organs affected. However, when you delve deeper into his writings, where he explains that diseases can also be caused by seeds in the body, and that these seeds are analogous to spirits that can be transmitted directly from one person's eye to another's, the connection to germ theory becomes somewhat tenuous. Perhaps we should view his writing more in light of the school of thought that characterized alchemy, in which seeds, in the larger sense of the word, meant anything that gives rise to something and has been scattered throughout the natural world by God.

Historical Overview

A Fascination with Dissections

Dissections and the resulting progress in anatomy are often seen as a victory for reason over ignorance and religious obstruction of the facts. In more recent times, François Jacob compared opposition to producing stem cells from embryos to the prohibition of dissection, as based on the same opposition to the spread of knowledge and refusal to accept its free use.[8]

7. Brock, 1999.
8. Jacob, 2000.

Nothing in this flawed history stands up to scrutiny. Dissection was in fact never prohibited for religious reasons, and, indeed, one has to question the science behind them when accounts give the impression that they were nothing more than spectacles. Furthermore, the progress in anatomy that they brought about had very little impact on medical practice over the next two centuries.

The beauty of the anatomic descriptions seems to have overshadowed their more practical application. This begs the question of whether this is an isolated case in the history of the biological sciences, or rather a phenomenon that repeats itself every time our attention is brought to bear on a new source of information. It must have been much the same with images from microscopes in the following century.[9] Two rather different examples can illustrate these risks. Putting together gene maps, which began in the early 1910s, quickly became a goal in and of itself, and meant that other pursuits were neglected, such as research into the chemical nature of genes or their mechanisms of action. Molecular biologists have also often been accused of practicing "molecular anatomy," a criticism that was justified when the detailed description of regulating sequences in promoters (or enhancers) took priority over research into a functional explanation of the system.

The Role of Alchemy

The importance of alchemy in the development of science has been largely ignored by contemporary historians.[10] However, it is well established that many notable scholars were adherents of the principles of alchemy (Newton and Boyle among them) and that alchemists played a role in the development of laboratories.

9. Freedberg, 2002.
10. Joly, 2013.

Despite their metaphysical leanings, alchemists no doubt facilitated the emergence of practical atomism, the precursor to atomic theory, through their focus on the purification of active principles. Far from the abstract approach of Greek atomists, their method paved the way for a modern atomism that sought to purify and describe the elemental components of matter.

By demonstrating that certain illnesses were caused by poisons found in the environment, Paracelsus spurred a radical change in our understanding of disease and the role of physicians. The causes of disease had become external rather than internal, and the focus of physicians' work shifted from the patient to the environment, thus laying the groundwork for the development of hygiene and the search for pathogens.

As we have seen, alchemists, and Paracelsus in particular, drew no distinction between the living and nonliving. To them, the Earth was a living being and living tissues obeyed the same laws of alchemy as minerals. A century later, Descartes and other practitioners of mechanistic medicine would also deny that there was any difference between living beings and inanimate objects, even going so far as building the first simple machines. These two opposing visions were united in their rejection of the uniqueness of living things. Contrary to the story that is often told, the development of vitalism at the beginning of the eighteenth century was not a rebirth of alchemy, but rather a break with both Descartes's mechanistic views and the ideas of alchemists.

Changes in the Social Structure of Science

The publication of works by naturalists was not a consequence only of the development of printing—these beautiful illustrated works were expensive to produce and naturalists were often not able to the cover the costs. It was only through the support of princes and kings that they were able to publish their works. This alliance between science and power would evolve over the next century, leading to the institutionalization of science through the establishment of academies.

However, this is not the only change in the way science was conducted that we can attribute to naturalists. The cabinets of curiosities that they created at home from samples they collected or received from other naturalists through trading networks that cropped up around the same time were the precursors of our museums of natural history. The role these institutions played in the development of scientific knowledge would reach its zenith in the nineteenth century.

Contemporary Relevance

Finding the Right Distance from the Past

It is very difficult when looking at the past to strike the right balance between the idea that past views were close to our own, or that our views perhaps even stemmed from them, and conversely that they were very different or even incommensurable, to use Thomas Kuhn's turn of phrase.[11] As we have seen, this challenge is very real when it comes to recognizing the contribution of alchemists.

Seeing the comparisons made by Belon between the human skeleton and bird skeletons as a first step toward comparative anatomy and the concept of evolution is one example of a backward-looking judgment, and a false one at that. The pervasiveness of evolutionary ideas is so strong today that any resem-

11. Kuhn, 2012.

blance to them is immediately seen as a sign of a relationship between the ideas. There was no such link for Belon, nor among the anatomists of antiquity who had compared human anatomy with that of animals. In fact, these scholars noted little more than obvious similarities in structure.[12] Conversely, Michel Foucault's thoughts on Aldrovandi are clearly too simplistic[13] Though Aldrovandi's views were acceptable to his contemporaries, other views and ways of presenting animals that chime with our modern sensibilities would have certainly been proffered by other naturalists from the same period. Foucault's famous fishbowl in which our ideas remain captive, today as in the past, should rather be viewed as an island from which we can escape, but at our own risk and peril.

Reading the writings of alchemists from the sixteenth century may be one of the most disorienting experiences there is, because of the confused jumble of disciplines that characterizes their texts. It would be easy to conclude, hastily perhaps, that this way of writing has no relevance to modern-day scientific writing. On the other hand, perhaps we could use this example to help identify the mixing of disciplines in contemporary publications—that is, the mixing of direct interpretation of experimentation with reflections that appear to be scientific but in fact amount to metaphysical speculation. Though this conjecture may be different from that expressed in the sixteenth century, it is no less rooted in metaphysics. This kind of thing can crop up when scientists use the criterion of simplicity to identify the "true" laws of nature. It also occurs when we give free reign to determinism (i.e., the notion that events are the result of external causes) or conversely to contingency (i.e., that

12. Schmitt, 2006.

13. Foucault, 1994.

events occur by chance) with respect to natural phenomena. Saying that the world is simple, determinate, or indeterminate is a statement that goes far beyond what the scientific facts show us, and at best is a hypothesis that remains to be tested.

New Techniques Bring New Sources of Error

While printing was an extraordinary tool for disseminating knowledge, it was also a very effective means of disseminating errors. When inaccuracies did arise, like in the representation of the rhinoceros, they were copied and amplified. This phenomenon, which is clear when comparing representations of the animal from different naturalists, is perhaps less obvious today, but is just as important in the information that scientists use in their work. It is well known that bibliographic references are often recopied without being read, leading to authors being credited with results that they never obtained or hypotheses that they never came up with. And computing tools are far worse when it comes to disseminating misinformation than the limited means of printing available in the sixteenth century.

Aging as a Form of Poisoning

Paracelsus was the first to provide an explanation of aging as poisoning, and similar theories would be embraced at least twice more in the history of biology.[14] At the beginning of the twentieth century, Élie Metchnikoff (1845–1916), who was known for his work in cellular immunology, proposed that aging was a result of the organism being poisoned by toxins produced by the bacteria found in the large intestine. He ar-

14. Morange, 2011b.

rived at this conclusion by comparing lifespan of animals and the relative size of their large intestine. As a remedy, he proposed a dairy-product-based diet, to replace the pathogenic intestinal microflora with a new microflora that would not produce toxins. The diet proposed by Metchnikoff and his theory were fairly quickly dismissed.

Today, a new theory of aging through poisoning is beginning to emerge. It differs from the earlier theory in that the poison is endogenous: aggregates of proteins form in cells and progressively disrupt cell function. Alzheimer's disease and Parkinson's disease, which are caused by such aggregates, are said to be specific examples of a more generalized phenomenon.

Other theories of aging exist, involving a loss of bodily functions and not, as in the preceding examples, the formation of toxic substances. It is curious to see that the same type of explanation has reappeared on at least three occasions, at different times, in slightly different forms. It is as though the "catalog" of hypotheses through which we view a phenomenon such as aging is limited.

4

The Age of Classicism (Seventeenth Century)

The Facts

The Discovery of Circulation

Based on a long series of prior anatomic observations, Harvey's 1628 work *De motu cordis* (On the motion of the heart), like Vesalius's work a century earlier, is considered to be a major contribution to the development of biological thought.[1] Indeed, historian Mirko Grmek went as far as to call it one of the events that shaped the first biological revolution.[2]

The model of the movement of the blood that dominated at the time had been developed by Galen in antiquity.[3] For simplicity's sake, we will make use of current nomenclature to explain the steps that made this discovery possible. The movement of blood, as Galen conceived it, was not akin to circulation, but rather like a sort of dual irrigation system. In his view, blood

1. Boustani, 2007.
2. Grmek, 1990.
3. Boylan, 2007.

was produced by the liver from nourishment coming from the intestine via the portal vein, and was then transported via the venae cavae toward other parts of the organism, where it carried nutrients, and also toward the right atrium and then the right ventricle of the heart. From there, a small portion made its way through the pulmonary artery, where some waste was eliminated from the blood. Most of blood in the right ventricle then crossed the interventricular septum via invisible pores, and once in the left ventricle, the blood mixed with the air carried by the pulmonary vein. Heated by the heart, it was then sent via the aorta throughout the whole organism, where it provided heat and vital spirit. In putting forward this model, Galen pitted himself against Praxagoras of Kos, a physician who believed that arteries carried only air from the lungs.

As we have seen, Ibn al-Nafis was the first, in the thirteenth century, to conceive of the notion of lesser circulation through the lungs. In his view, blood did not cross the interventricular septum of the heart but rather passed through the lungs, where it absorbed air.

Ibn al-Nafis put forward two lines of reasoning to support the new model: there were no pores in the interventricular septum and the pulmonary vein contains a mixture of blood and air rather than simply air (which was already known by Galen, but he interpreted it as being blood reflux from the left ventricle). Indeed, it is wrong to speak of lesser circulation, as it is the lesser and greater circulation together that constitute circulation in the true sense of the word.

The valves in veins, which facilitate the return of blood toward the heart, were described by Erasistratus in Alexandria in the third century BCE. However, this observation had been forgotten, and they were later redescribed by anatomist Fabricius ab Aquapendente, whose work we have already described.

Fabricius ab Aquapendente's interpretation of the valves was flawed, however—he believed them to be regulators or locks in the organism's blood irrigation process.

Three centuries after Ibn al-Nafis, Michael Servetus (1511–1553), a physician from Spain, would independently advance the hypothesis that blood did not cross the interventricular septum, but rather passed through the lungs. In addition to the arguments put forward by Ibn al-Nafis, he remarked that the large size of the pulmonary artery, comparable to that of the aorta, did not square well with the role attributed to it of eliminating waste via the lungs. His hypothesis went almost as unnoticed as that proposed by Ibn al-Nafis three centuries earlier, as he included it in a theological work. Indeed, it was not his revolutionary ideas on the movement of blood that would lead to him being burned alive in Geneva under orders from Calvin, but his criticism of the dogma of the Trinity.

New observations progressively undermined Galen's model. Vesalius did not contribute to this process in a significant way. Even though, like other anatomists, he could not see any pores in the interventricular septum, he imagined them to be there nonetheless, and it would not be until the 1555 edition of his work that he would be more circumspect in his statement of the existence of these invisible pores.

In 1542, Jean Fernel (1497–1558), a physician who remained faithful to Galen's work and who is known for having been the first person to use the term "physiology" in its modern sense, showed that, contrary to what had been previously believed, the heartbeat did not result from expansion of the heart but rather from its contraction. Vesalius's successor as chair at Padua, Colombo, confirmed this observation and also supported the position that blood passed through the lungs as he found only blood when he opened up pulmonary veins. To the arguments already put forward, he added that the perfect functioning of the mitral valve prevented any reflux of blood from the left ventricle into the pulmonary vein. Another explanation for the presence of blood in this vein was therefore needed. Andrea Cesalpino introduced the term "circulation of the blood," though he did not seem to have had a clear idea of the cyclical movement of blood in the organism.

However, Paré and Falloppio were opposed to this new view. The latter drew from direct observations of the fetus in which he noted the absence of pulmonary circulation, which is partially true. What he didn't know is that in the fetus a conduit between the two atria and another between the pulmonary artery and the aorta allow the blood to bypass the lungs, but these conduits disappear after birth.

William Harvey (1578–1657) studied at Cambridge, then made his way to Padua, where he was taught by Fabricius ab Aquapendente. After his return to England, he was appointed to the office of Lumleian lecturer in anatomy in 1615, and soon after physician to King James I and later Charles.

He came up with the idea of blood circulation through the action of the heart as a "pump" in 1616. His 1628 work, whose complete title is *Exercitatio anatomica de motu cordis et sanguinis in animalibus* (*An Anatomical Disquisition on the Motion of the Heart and Blood in Animals*), comprises 17 chapters. The first 7 chapters precisely describe the motion of the heart, the eighth puts forward his hypothesis of circulation, and the remaining chapters outline the arguments in favor of this hypothesis.

At least two factors contributed to the impact of the book—the clarity and simplicity of Harvey's style of writing and the experiments supporting his model. These experiments were carried out on several species of animals, including amphibians, reptiles, and fish. Each of these organisms was chosen for a different reason, to make studying the motion of the heart easier. The hearts of certain species have simpler structures, or beat more slowly, which allowed Harvey to distinguish the contractions of the atria and the ventricles. In other cases, the organism was translucent, revealing the motion of the heart. Harvey is not the only one to have experimented on live animals, what we call vivisection. For anatomists, it was a return to a forgotten practice from antiquity. However, as we have seen, Harvey had an inventive spirit. His most memorable experiment was to estimate the quantity of blood that passed through the human heart in an hour. Though he underestimated it by a factor of 30, the figure obtained by Harvey demonstrated that it was not feasible for the blood to be created anew by the organism, but that it must rather be recycled via circulation. Harvey also described the different effects that resulted from ligaturing veins and arteries, both in humans and animals, and drew some conclusions on ways to conduct bloodletting. His experiments confirmed the role played by valves in veins. He demonstrated that his model could offer the first explanations of certain phenomena, including the rapid spread of poison in organisms, but also diseases such as rabies that resulted from the bite of a rabid dog. It also explained how a medication applied to the skin could rapidly have an effect on an organ some distance from the site of application.

However, Harvey's model is less perfect than it may appear. The reason why blood makes a detour to the lungs is not explained. Harvey envisaged the flow of blood from arteries to veins, but could not see it happening with the simple magnifying glasses he was using. Capillaries would only be observed under the microscope by Marcello Malpighi in 1661.

However, it was not for these reasons that Harvey's model was criticized by Parisian physician Jean Riolan (1580–1657), considered one of Europe's foremost anatomists, or by philosopher

Pierre Gassendi (1592–1655).[4] Rather, it was their attachment to Galen's theory, an attachment that was very strong at the faculty of medicine in Paris, which partly explains their opposition. It was also due to a shared mistrust of vivisection, which revealed phenomena occurring in conditions that are far from natural—a criticism that had been made by Roman physician Celsus in the first century. In addition, they criticized Harvey for not showing the causes behind the phenomena that he observed, and in particular the contractions of the heart—even though Harvey demonstrated that the heart had the structure of a muscle.

If Harvey was revolutionary in his scientific thinking, these revolutionary ideas did not extend into the philosophical or political spheres. He adhered to Aristotle's view of the heart as a central part of the organism, and for this reason, the first organ to be formed during the development of the embryo. Like Aristotle and Galen before him, he subscribed to the finalistic view that natural processes were directed toward an end goal—the most oft-repeated phrase in his books is that "Nature does nothing in vain." His model is in keeping with the notion of a correspondence between the macrocosm and the microcosm—the circular movement of blood in the organism reflects the circular movement of the stars. Harvey compared the central and beneficial role of the heart to the sovereign's role in society. He was also opposed to the atomist ideas of some of his contemporaries.

The Development of Quantitative Experiments

Santorio Santorio (1561–1636) is no doubt the best example of the new breed of scholars who introduced, or, more precisely, reintroduced (after the Alexandrians), quantitative experiments

4. Guerrini, 2013.

in biology.[5] They applied or even anticipated Galileo's famous phrase "measure what is measurable, and make measurable what is not so."[6] Born in Istria (Slovenia), Santorio made his way to Venice and then Padua, where he published his first work in 1602, on a method for avoiding errors in the healing arts. He was appointed professor of theoretical medicine in 1611. He was a friend of Fabricius ab Aquapendente and knew Galileo well, but was not one of his disciples. In 1614, he published *De statica medicina* (On medical measurement), in which he described the instruments that he had developed. The first was a pendulum that he used to measure the pulse and variations in it—essentially reversing Galileo's experiment where he used his pulse to measure the speed at which a pendulum oscillated. The second was a thermometer based on the expansion of liquid (first water, then alcohol) with rising temperatures, which Galileo is often wrongly credited with inventing. Santorio soon made a key change to the first thermometers by introducing regular graduations between the temperature at which water freezes and that at which it boils. Following in the footsteps of Nicholas of Cusa (1401–1464) and Leonardo da Vinci, he built a hygrometer to measure the level of humidity in the air and an anemometer to estimate the speed of the wind. Finally, he created scales adapted to weigh a whole human being in order to quantitatively estimate the body's "invisible transpiration," a question that had remained unanswered since antiquity. These scales would allow him to demonstrate that the weight of the human body does not change during sleep, and that cadavers weigh the same as living bodies.

As much as we may admire Santorio's inventions, there are two caveats that are worth noting. The first is that his instruments

5. Grmek, 1990.

6. Quoted in Grmek (1990, p. 73)

allowed measurement of the qualities of the environment that since Hippocrates had been believed to be essential to human health—and thus Santorio did not break with the medical tradition of antiquity. The second is that his inventions did not have an immediate impact—the thermometer would make its way into common medical practice only in the eighteenth century. These are two serious limitations of his work, which otherwise seems so modern to us.

In the year of his death, Giovanni Alfonso Borelli (1608–1679) published a work titled *De motu animalium* (*On the Movement of Animals*), the result of work carried out at the Accademia del Cimento in Florence, in which he described the role of muscles in movement, both in humans and in animals. Loyal to Descartes's method, he first depicted simpler movements, before moving on to more complex ones like swimming. His mechanical ideas could not, however, explain where the driving force behind these movements came from, which he attributed to fermentation resulting from the mixing of the blood and nervous fluid. Borelli also directly measured the heart temperature of a live doe—a spectacular and bloody experiment to examine a fundamental issue. Since Aristotle, through Galen and Harvey, and Descartes's fire without flames, it had been accepted that the heart was the organism's "boiler," or source of heat. Borelli demonstrated that the temperature of the heart is the same as that of other parts of the organism, delivering the death blow to this view.

In a similar vein of spectacular and bloody experimentation, we can include later measurements of blood pressure in various animals by Stephen Hales (1677–1761). Remarkably, the figures he came up with are very close to today's accepted values.

At the same time as Borelli, Claude Perrault (1613–1688) was independently carrying out analogous work at the French

Academy of Sciences in Paris, which he described in the third part of his four-volume *Essais de physique* (Essays on physics), titled *Mécanique des animaux* (Animal mechanics), published in 1680. Perrault is better known for his architectural work (he was involved in the restoration of the Louvre) and for his support of the Modernist school in architecture. The analogies that he used to explain the genesis of movement in animals were borrowed directly from machines of the period. When it came to philosophy, he opposed Descartes's ideas and warmed to those of Pierre Gassendi, and he used the argument that animals have souls to explain their movement, while drawing on an atomistic view of matter.

Italian physician Giorgio Baglivi (1668–1707) would call this new approach to living phenomena "iatrophysics." In this approach, the organism is compared to a machine made up of levers and springs, whose operation can be explained in physical terms, as opposed to chemical explanations, involving ferments, which Baglivi renamed "iatrochemistry."

It is difficult to position the work of Descartes (1596–1650) in the context of the work being carried out during the period. Galileo's influence in the development of biophysics and quantitative biology was as strong as that of Descartes, and the latter's concept of the "animal-machine" was far from being unanimously accepted among naturalists. Indeed, as we have seen, it was rejected by Perrault, who gave animals souls while still maintaining a mechanistic approach to the living world. Descartes carried out numerous dissections while living in Amsterdam, from 1630 to 1631. He set out his ideas on human physiology in *Les passions de l'âme* (*Passions of the Soul*), *L'Homme* (*Treatise of Man*),[7] and *Premières pensées sur la génération des animaux* (First thoughts con-

7. Descartes, 2003.

cerning the generation of animals). Unfortunately, his physiological work, like his work in physics, contains many errors.

His physiological work drew heavily from Galen, and despite accepting Harvey's model, he distorted it by making the heart's dilation the active phase in its motion, rather than its contraction. The idea that nerves are home to animal spirits comes from Galen—the only change Descartes made was to make the pineal gland the seat of the soul, rather than the third ventricle. Suspended at the base of the brain, the pineal gland could be set in motion by tiny impulses, which meant Descartes didn't need to abandon his distinction between thinking matter and expansive matter (which is capable of causing motion).

What should biologists think about Descartes's work? Should they see it as the starting point for the body of research that would follow in later centuries? Should they remember him for the remarkable ways in which he anticipated future developments, such as his description of the projection of images in the brain? Similarly, did Descartes not anticipate the genetic program of embryonic development when he wrote that if we understood all of the parts of an organism's reproductive seed, we could deduce from it "the whole conformation and figure of each of its members"?[8] However, biologists might also be tempted to consider Descartes's texts as pure fiction attempting to compensate for a total lack of knowledge, both on the origins of physiological processes and to an even greater extent on mechanisms of embryonic development.

Jan Swammerdam (1637–1680) soon showed Descartes's model of the functioning of nerves to be false.[9] If nerves were hollow structures that transported animal spirits, as Galen and

8. Descartes, 2000, p. 277.

9. Cobb, 2002.

Descartes believed, then neural activation should make muscles expand. Swammerdam worked with isolated frog nerve-muscle preparations, which he placed in a liquid to allow him to measure the volume of the muscle before and after stimulation by the nerve. Contrary to what we would expect, the experiment showed that the muscle's volume decreased when it contracted. Beyond demonstrating that there is no nerve fluid (of any kind), this experiment demonstrated how we can artificially stimulate a nerve and observe the effects in a simple experimental system. This nerve-muscle preparation would be used widely and often in the centuries that followed.

What was more important in the seventeenth century was not so much the growing influence of mechanistic models as the real progress made in descriptive anatomy. From the spectacle that it had been in the sixteenth century, anatomy would become a science built on cumulative knowledge where reliable observations accumulated independently, with no preconceived system. Perrault was an anatomist much more than a mechanist—in fact, he died of an infection caught while dissecting a camel.

The discovery of the lymphatic system is a perfect example of real progress. The first lymphatic vessels, which were called lymphatic veins, were described by Gasparo Aselli (1581–1626), a professor at the University of Pavia, who suggested that they transported the end products of digestion from the intestine to the liver. He reached this conclusion in 1627, at the same time as Harvey was publishing his work, and thereby interfering with Harvey's findings. His description of these lymphatic vessels reinforced the idea that blood was produced in the liver (although it challenged the notion that food passed from the intestine to the liver via the portal vein, which had been the accepted view since antiquity), and seemed to go against Harvey's hypothesis that blood was recycled through general circulation.

Jean Pecquet (1622–1674) in Montpellier, Thomas Bartholin (1616–1680) in Copenhagen, and Olof Rudbeck (1630–1702) in Uppsala are credited with making the lymphatic system a new circulatory system. More specifically, Pecquet showed that lymphatic vessels did not go directly toward the liver but rather joined up with general blood circulation via the thoracic duct.

We must also mention the work carried out in London by Thomas Wharton (1614–1673) and Thomas Willis (1621–1675). The former described the nature of glands and the latter made considerable progress in anatomic descriptions of the brain using a comparative approach, ascribing different cognitive functions to its main structures.

The Invention of the Microscope and Its Consequences

The first microscopes were built at the end of the sixteenth century and they were improved as their use spread, over the seventeenth century. The first book of micrographs was published in 1625 by Francesco Stelluti (1577–1652), and showcased beautiful images of the heads and mouthparts of bees, in particular. There are four authors who are worthy of our attention in this area. Robert Hooke (1635–1703), one of the founders of the Royal Society in London, published his work *Micrographia* in 1665, in which he presented several magnified representations of insects and arachnids, but also birds' feathers and plant tissues such as cork. He was the first to observe the regular structures in cork that he called "cells," referring to the cells of a monastery. Considered more important owing to the quality and range of their observations were Jan Swammerdam and Antonie van Leeuwenhoek (1632–1723). Swammerdam was the first to describe red blood cells, in 1658. He is best known for his magnificent microscopic images of insects, including lice. Micrographic representations benefited from the growing use of copperplate engraving, which yields a much higher resolution image than wood plate engraving. The historian Jules Michelet has called Swammerdam the "Galileo of the infinitely small."[10] Swammerdam gave a religious dimension to his description of the microscopic world: "I offer you the Omnipotent Finger of God in the anatomy of a louse."[11] He described the metamorphosis of insects in meticulous detail and showed the organism's persistence during this transformation. He highlighted the ovaries present in the queen bee, but remained unable to describe the mechanism of fertilization.

It was Van Leeuwenhoek, a draper from Delft, who perfected the loupes used to examine fabrics.[12] His work was unique in that he did not use a compound microscope, but rather magnifying glasses—a technology that had been around for centuries, which he improved to such an extent as to rival the new microscopes. He built more than 300 small magnifying glasses over his lifetime, which allowed him to reach a magnification of 250 times and to discern objects 0.0015 millimeters apart. The way in which he lit his samples remains a mystery. With these tools, he observed everything that was at his disposal. He sent more than 300 letters to the Royal Society in London, which would be translated and published by its secretary, Henry Oldenburg (*ca.* 1619–1677), in the society's *Philosophical Transactions*. He described red blood cells and spermatozoa (which he called animalcules), protozoa, and the bacteria found both in stagnant water and in dental tartar.[13]

10. Michelet, 1858, p. 92.

11. Quoted in Enenkel and Smith (2007, p. 63).

12. Boutibonnes, 1994.

13. Berche, 2007.

However, it is Marcello Malpighi (1628–1694) who is considered to be the founder of microscopic anatomy.[14] He improved techniques for preparing samples for microscopic observation. As we have seen, he described capillaries, providing confirmation of Harvey's theory, and the fine structure of the cerebral cortex, taste buds on the tongue, and the spleen. He also turned his attention to studying insects and wrote a treatise on the anatomy of the silkworm (1669). His main contribution is to have expanded the mechanistic view of the living world to the microscopic level.

Malpighi also conducted work on plant anatomy, using observations carried out on animals and, in particular, insects as a guide. He drew parallels between bone tissue in animals and woody tissues in plants.

At the same time, English anatomist and physiologist Nehemiah Grew (1641–1712) also described the microscopic anatomy of plants. He discovered the vascular system in plants independently from Malpighi, but allowed the latter to take credit for the discovery. Most of his observations, and in particular his descriptions of pollen, are collected in his work *The Anatomy of Plants*, published in 1682.

The most important consequence of the use of microscopes was that it called into question models of reproduction passed down from antiquity. These involved the mixing of reproductive fluids or, for Aristotle, the animation of the female reproductive seed by the male reproductive fluid, which was followed by the progressive building of the organism (said to be a model of epigenesis, after Harvey described it in this way).[15] Following the discovery of animalcules, or microscopic animals, it was the discovery of eggs in both insects and mammals that promoted the

14. Bertoloni Meli, 2011.

15. Bowler, 1971; Cobb, 2006.

development of the theory of preformationism. The Dutch Regnier de Graaf (1641–1673) described what he called eggs in the ovaries, which would later be shown to be ovarian follicles, of which eggs are only a minor part. The first rudiments of the theory of preformationism were found in the study of Venetian physician Giuseppe degli Aromatari (1587–1660) on plant reproduction. Harvey observed early embryos that were already well formed in a doe uterus, and Nicolas Steno (1638–1686) made the same types of observations in a shark uterus. These observations could be interpreted as the rapid metamorphosis of the egg after fertilization; however, the notion that the organisms is already formed in the animalcule or the egg, and only grows over the course of embryogenesis, would gradually take hold. Dutch scholar Nicolaas Hartsoeker (1656–1725) drew in 1694 the small, crouching human being that he expected to see in the heads of spermatozoa.

The theory of emboîtement, or encasement, of germ cells was gradually added to preformation theory, whether based on eggs or animalcules, and was supported by philosopher Nicolas Malebranche (1638–1715). According to this theory, not only is the organism preformed, but it already contains animalcules or eggs in its testes or ovaries, which themselves contain the animalcules and eggs necessary to produce future generations.

In 1668, Italian naturalist Francesco Redi (1626–1697) showed in his *Esperienze intorno alla generazione degl'insetti* (Experiments on the generation of insects) that worms (and flies) were not spontaneously generated from decomposing meat, but rather from eggs laid by the flies. When Redi used a screen to prevent the worms from getting to the meat, it did not prevent the meat from rotting, but no new worms developed. Redi's experiment separated the living from the nonliving and showed life's continuity. In the absence of any ideas on the persistence of cellular forms, the results of his experiment were more in keeping with the theory of preformation than that of epigenesis.

Historical Overview

The Not-So-Obvious Case of Circulation

The slow progress in developing a model of circulation cannot be explained by the obstacle that Galen's theory represented, but is merely because the phenomenon itself is not obvious and observations were rather ambiguous. Understanding the rapid movements of the heart in a living animal, even an "open" one, is problematic at best. The folding of veins and arteries near the heart does not make the function of this pump easy to decipher. Precise observations were conducted many times, but were not interpreted or were misinterpreted. The presence of air in the arteries was not an illusion—it is a phenomenon that occurs soon after the heart stops beating. But this does not mean that the physiological role of the arteries is to transport air. Likewise, the difference in color between arterial and venous blood was observed very early on, but its significance would remain unknown. We have also mentioned the precise observations of the fetus that were conducted, but which would prove counterproductive. We should not interpret all discoveries that appear to have been difficult as overcoming some sort of obstacle in knowledge theory, or as getting out of an "epistemic fishbowl." Sometimes the difficulty is simply due to the obscurity of the phenomenon.

The Mechanistic Model of Life and Its Limitations

The seventeenth century is considered the golden age when it comes to mechanistic views of biology.[16] Perrault, Descartes, and Baglivi were constantly comparing living organisms to machines. A mechanistic view of the living world does not, however, necessarily mean comparing them in this way—it can

16. Canguilhem, 1965b.

simply involve stating that organisms are subject to the same physical principles in their functioning as are machines. These new explanations of living phenomena can be called mechanistic only by extension, because they draw on physical explanations and physics is considered to be mechanistic.

As an example, in 1638, Galileo proposed a new explanation of "plant diseases."[17] These diseases had been known since antiquity and studied in great detail because of the famines they caused. Since Theophrastus, it had been believed that they were a result of the successive effects of fog and the sun, which caused the decomposition and rotting of the fluid found on the surface of plants. Galileo proposed a more "physical" model: fog droplets were perfect spheres that focused light from the sun on fine points on the surface of leaves, which caused burns. Galileo's explanation is in no way a comparison of organisms to machines—but it is a purely physical explanation that does not draw on characteristics unique to living things. This is the second principle that underpinned the mechanistic explanations that we typically come across in the seventeenth century.

In addition to the ambiguity of the term, mechanistic and chemical explanations involving the action of ferments often coexisted in the thinking of many scholars, as was the case with Descartes as well as Perrault and Malpighi. While mechanistic explanations have their strong points, known since antiquity, they also have their limitations. While it is possible to describe the action of muscles in purely mechanistic terms, it is more difficult to do the same for digestion, which would remain the privileged domain of alchemists and, later, chemists.

Moreover, we should not overestimate the importance of mechanistic models in the development of new knowledge.

17. Denis, 2011.

While Harvey mentions that the heart is a pump, he does not seem to give much importance to this analogy. What was more significant for him was the heart's central place, which had already been highlighted by Aristotle, analogous to that of the sun in the universe or the sovereign in society.

The Incomprehensible Theory of Preformationism

Preformation theory and the theory of the encasement of germ cells appear to be historical aberrations. For a century, they replaced the theory of epigenesis, which had been proposed in antiquity, and which still forms the foundation of our understanding of embryonic development today. How do we explain this lapse? Religious reasons are often put forward: that preformation theory could justify how God was able to create all living beings—past, present, and future—at the moment of the creation of the world. It meant the Creator did not need to intervene in the creation of each new living creature, which was more in keeping with His role in establishing the laws of nature attributed to Him by the new physics. However, it was the theory of encasement of germ cells, and not preformation theory, that provided the solution to this theological problem; and it would have seemed quite foreign to the first adherents of preformationism.

Perhaps we should rather see the success of preformationism as a direct consequence of the development of scientific knowledge. The shock of the discovery of the microscopic living world surely led naturally to conjuring up worlds on an infinite scale. Echoes of these ideas can also be found in Pascal's and Descartes's work. The notion of the infinite divisibility of matter, which seems absurd to us, as we know that the atoms and molecules that make up living things have a precise size, would not have seemed so absurd in the seventeenth century, even for those who were partial to atomist ideas (the size of atoms would remain undefined). The success of preformation theory was also a result of the failure of Descartes's project, which sought to describe the development of organisms in mechanistic terms. Descartes's fabricated ideas would therefore no longer hold sway. Many scholars have noted what appears to be an internal contradiction in Descartes's work: the unique quality of a machine is that it is put together by the person who builds it, who is external to it. It is not plausible to compare an organism to a machine, while at the same time having this organism be capable of building itself. Preformation theory allows us to keep this analogy between organisms and machines, without having to deal with the thorny issue of how they were formed.

Invisible and Indirect Changes

The seventeenth century was a crucial period during which other changes, some invisible and some indirect, would come about, which would contribute to ensuring the future development of biological knowledge. The first was the difficult and progressive renunciation of the main systems of philosophy in favor of a more modest, but also cumulative, vision of scientific work. Scientists were not spared in this upheaval of ideas and frameworks of thought. For some, like Swammerdam and Steno, mystical crises interrupted their scientific work. Changes in scientific ideas, and even more so in metaphysical frameworks, are always met with great resistance. Nonetheless, the landscape was changing. Anatomy would progressively cease to be a spectacle revealing the Creator's greatness and the wisdom of the ancients, and become a commonplace field of study, where observations that were made would patiently await explanation.

Herman Boerhaave (1668–1738), a physiologist working in Leiden, was typical of this shift. He was unanimously honored for his synthesis of anatomic and physiological knowledge, the importance that he gave to mechanistic explanations, and also his restraint in looking for the root causes of phenomena.

Another, more important but equally silent, transformation occurred in scientific instruments and techniques, resulting in the development of the experimental method. The increasing use of microscopes is one example, but there were also changes and improvements to many other instruments used in anatomic and physiological study. Although this progress may be difficult to perceive, it can nonetheless be gleaned from careful study of works on experimental devices used by scholars from the period. This progress ran parallel to the development of small scientific instrument companies that specifically targeted scholars.

The third change was the establishment of scientific academies. While the creation of the Accademia dei Lincei, or Lincean Academy, in Rome in 1603 was precocious, the main European academies of science would be established over a relatively short period between 1650 and 1670: Leopoldina in Halle, Germany, in 1652; the Accademia del Cimento in Florence in 1657; the Royal Society in London in 1661; and the French Academy of Sciences in Paris in 1666.[18]

Academies broke with universities and their ways of teaching. They consolidated the scientific networks that had been built up from the sixteenth century, and the financial support of princes and kings. Academies determined the rules by which the results of experiments and observations were to be communicated—through the creation of scientific journals like the *Philosophical Transactions* in London and the reports from the French Academy

18. Salomon-Bayet, 1978.

of Sciences in Paris. They also set the ethical rules by which members of the scientific community were bound. Thus, they were a part of the movement that provided stability in science that we have previously described. When it came to the life sciences, they also recognized some separation of the discipline from medicine and contributed to reinforcing the importance of physics- and mathematics-based approaches in biology.

Contemporary Relevance

The Machines in Front of Us

When they compared the human body or certain body parts to machines, scholars from the seventeenth century looked to the machines that they had in front of them as points of reference. Descartes used the analogy of the organ (the musical instrument), a machine that was commonplace.

This seems to be a general rule in science. At the beginning to twentieth century, Charles Sherrington compared the organization of nerve connections to telephone line connections. The models used in science are borrowed from one's immediate surroundings and from current affairs—a rule that applies as much in our time as any other. Machines no longer feature in the news today, but rather computers and, even more so, communication networks. Is it purely a coincidence that networks feature so heavily in the work of cellular and molecular biologists today?

Vestiges of Preformation Theory

Modern biologists consider preformation theory to have been an aberration. However, are there not whiffs of a kind of preformationism in considering a particular variant of a gene to be responsible for specific psychological characteristics or behaviors? In general, genetic determinism can be considered a manifestation of preformationism. We will see as well that the criticism that genetics was a return to preformationism was leveled from the very start.

Accepting the Plurality of Approaches in Biology

In the seventeenth century, many interpretations of living phenomena clashed with one another. Baglivi distinguished approaches based on iatrophysics, which he supported, and ones based on iatrochemistry, inherited from alchemy, which he condemned as obscure. We have seen how in practice it is difficult, if not impossible, to avoid chemical explanations in describing digestion or the progressive development of the embryo.

The same argument seems to go on today between proponents of molecular biological approaches—that is, chemical approaches—and those who defend physics-based methods for examining living phenomena.

Conflicts such as this most certainly arise from the hegemonic tendencies of each discipline, whose supporters are unable to admit that a combination of different explanations can often bring about a better understanding of a phenomenon like, for example, digestion. In the seventeenth century, as is the case today, scholars also tended to view the characteristics of the science they practiced as common and necessary to science as a whole. For example, the difficulty in establishing laws in the field of biology is still often considered by physicists a sign of how far biology has fallen behind, as was the case in the seventeenth century.

Translational Medicine Is Not New

Every important biological discovery should lead to applications in the medical field. The tools of genetic engineering, stem cells, and interfering RNA were immediately incorporated into projects to develop medical treatments, with varying degrees of success. This is all the more evident today with translational medicine, whose stated goal is to facilitate turning basic discoveries into medical applications.

However, this was also the case in the seventeenth century. After Harvey's discovery, transfusions of animal blood were conceived as a way of curing certain human diseases, taking the place formerly occupied by bloodletting.[19] Though the French Academy of Sciences did not look on it favorably, Jean-Baptiste Denis (1640–1704), the famous physician to Louis XIV, made many attempts at it, the first of which had some success. However, in 1668, he transferred blood from a calf to a servant, Antoine Mauroy, who was suffering from anxiety. Mauroy died during the third transfusion. Although Denis was not convicted during the trial that followed, the French parliament decided to ban blood transfusions the following year. Transfusions of blood between animals and humans were also carried out in England. In retrospect, it is almost a miracle that these individuals were able to survive the treatments. Blood transfusions between humans were conducted again in the nineteenth century to try to treat severe blood loss. The results, once again, must have been mixed.

19. Tucker, 2011; Bertoloni Meli, 2011.

5

The Enlightenment (Eighteenth Century)

The Facts

Vitalism

Few scientists are as universally derided in traditional historiography as German physician Georg Ernst Stahl (1660–1734), who is responsible for two false theories: one in chemistry (phlogiston theory) and one in biology (vitalism).[1] Phlogiston theory came about by combining observations on the transformation of metals that were rooted in alchemy and an understanding of combustion reactions. The theory postulated that, when exposed to air, metals released an element known as phlogiston, which was also released by other forms of matter during combustion. It was said that if phlogiston were added to a metal, it would regain its former qualities. Plants were thought to capture phlogiston, and the animals who consumed them would use it and then release it. By the end of the century,

1. Canguilhem, 1965a; Reill, 2005; Nouvel, 2011; Normandin and Wolfe, 2013.

Lavoisier had demonstrated that the supposed loss of phlogiston coincided with an increase in weight, thus disproving the model proposed by Stahl. He replaced the release of phlogiston with oxidation.

Sthal's second mistake in setting out his vitalistic notions was to have reintroduced the role of the soul to explain living phenomena. His work in this area stemmed from his criticism of the Descartes school for having denied the very existence of life. Reintroducing the soul as the source of life in organisms was a way of restoring the special status of living beings. Stahl nonetheless gave the soul a new function: to protect the body from the corruption that perpetually threatened living beings by virtue of their structural complexity. While he was familiar with alchemy, Stahl distanced himself from this tradition by separating the living world from the inanimate world.

Stahl's physiological work was not terribly original. Like Paracelsus, however, he subscribed to the idea that thoughts can cause diseases. But he did not exclude mechanistic explanations from his physiological work.

Stahl's writing is difficult to decipher, which also did not help his reputation. Nonetheless, some observers have noted that his phlogiston theory could easily be substituted with oxidation theory, which led some to see Stahl as the father of oxidation-reduction and of the modern notion of energy exchange in the living world. Stahl's spiritualism is more difficult to account for, even if it was an understandable reaction to the misuse of mechanistic models of life, which were often ill adapted to their subject.

Of more interest is the form of vitalism that would take root in the medical school at Montpellier. Through his work on glands, Théophile de Bordeu (1722–1766) demonstrated the inadequacy of models based on iatrophysics, in which secretion from glands was thought to be the result of increased pressure from the contraction of the muscles surrounding them. Bordeu showed that this was not the case and that secretion was initiated by nerve action that increased blood flow in the glands. This is a good example of a physical explanation proposed too quickly that, like that of Galileo to explain the origin of plant diseases, fails to recognize the complexity of living phenomena. Bordeu was a well-known scholar of the period—he is one of the people who appears in *D'Alembert's Dream* (1769), a collection of three philosophical dialogues by Denis Diderot (1713–1784).

Paul-Joseph Barthez (1734–1806) did not content himself, as Bordeu had done, with exposing the weaknesses of certain mechanistic models. He examined the causes of living phenomena, which he attributed to the existence of a vital principle in *Nouveaux éléments de la science de l'homme* (New features in the science of man), published in 1778. This vital principle was not a spirit—Barthez did not adopt Stahl's spiritualist ideas any more than Bordeu did. However, he believed in the existence of this vital principle, whose nature he did not question, as the only way of accounting for the properties that characterize living things, which were irritability (of muscles) and sensitivity (of nerves). An analogy is often made between this vital principle and Newton's universal gravitation, though the origins of neither could be explained. However, conceding that these principles exist can help us explain living phenomena in the case of the former, and phenomena of the inanimate world with the latter. Bordeu and Barthez both contributed to the French *Encyclopédie*, which played a part in the influence of vitalism on Enlightenment thinking.

The Montpellier school was hesitant when it came to conducting animal experiments. As far as its adherents were concerned, life could be studied only within its natural limits. Indeed, this "experimental abstention" was one of the charges laid in the case against vitalism. The Montpellier school also advocated a return to Hippocratic medical theory and respect for life's self-healing force. According to this theory, physicians should not intervene in an aggressive way—as in the example of blood transfusion—but rather treat patients by supporting the healing work of nature. Given the gaps in medical knowledge at the time, this prudence was no doubt more beneficial to patients than certain forms of more active intervention.

This "soft" form of vitalism also influenced the work of physiologist Albrecht von Haller, as well as embryologist Caspar Friedrich Wolff. It was the origin of Bichat's ideas and would endure for much of the nineteenth century among many biologists, including Claude Bernard.

Classification: Linnaeus versus Buffon

The classification of plants and animals remained relatively unchanged during the sixteenth and seventeenth centuries and continued to be based largely on the principles set out by Aristotle and Theophrastus. As we have seen, Andrea Cesalpino was the first to propose a classification based on the shapes of flowers and fruit and the number of seeds.

Joseph Pitton de Tournefort (1656–1708) continued in much the same vein, basing his classifications on flowers and fruit, as he believed these characteristics remained the most consistent within a species. He would add classes to the taxonomic categories of species and genera developed by Aristotle.

John Ray (1627–1705) gave up a position at Cambridge when he refused to sign the Act of Uniformity, as required by King Charles II, not because he subscribed to the ideas of those who fought against the king, but rather because he believed that the act would undermine his freedom of conscience. It was therefore as an amateur that he published his 2800-page folio *History of Plants*, which was a systematic review of all known plants. He gave greater importance to the shapes of flowers and fruit when developing his classification, though he also believed that the shapes of leaves and other characteristics should not be neglected. Vigorous debate with Tournefort would ensue, somewhat reminiscent of Buffon's opposition to Linnaeus's classification nearly a century later.[2]

Ray introduced certain concepts that would later be recognized as important—such as the division of plants into monocotyledons and dicotyledons (cotyledons are the embryonic leaves that grow out from the seed)—which he used as the basis for his classification. He also gave thought to the notion of the species, which he defined as the ensemble of organisms able to reproduce together, an idea that would later be taken up by Buffon. He acknowledged that there could be some deterioration in a species, which could even result in significant transformations of organisms.

The publication of *De sexu plantarum* (On the sex of plants) by German physician Rudolf Jakob Camerarius (1665–1721) reinforced the idea that a natural classification of plants should stem from studying the reproductive organs. While sexual differences were already known in plants and used in agriculture—artificial pollination of date palms had been practiced since antiquity–the fact that the majority of plants are hermaphrodites deterred naturalists from taking interest in their mechanisms of reproduction. Camerarius showed through experiments that flowers contained the sexual organs, and that pollen is the male part carried by the stamens that fertilizes the pistils

2. Sloane, 1972.

of female flowers. His demonstration would justify basing the classification on the shape of flowers, because reproduction is what defines a species, as we have seen.

Carolus Linnaeus (1707–1778) had an interest in botany, inherited from his father, from a very young age. After medical studies at Uppsala, he traveled to Lapland in northern Sweden, and wrote a very successful report about the trip. He then spent several years in Holland, where he published the first edition of *Systema naturae* (The system of nature). He was made a professor at Uppsala in 1741, where he organized botanical outings that drew more than 300 participants. He gathered a community of student enthusiasts around him that he would send around the world—including North and South America and Asia (China and Japan)—to complete the classification that he had undertaken. He called his students his "apostles." He published many works, including his major work *Species plantarum* (Species of plants) of 1753. After his death, his library, manuscripts, correspondence, and collections of samples (more than 14,000 plants and 3000 insects) were bought by the Linnean Society in London, where they are preserved to this day.

Linnaeus made classification his life's goal, and it became an obsession. He would famously boast: "God created, but Linnaeus organized."[3] He was also considered the "new Adam," as he was continuing Adam's work naming plants and animals as they were described in Genesis.

Linnaeus's main contribution was not his classification of plants, which would later be abandoned for the classification proposed by Bernard de Jussieu (1699–1777), published by his nephew Antoine-Laurent de Jussieu (1748–1836), and formalized by Augustin Pyrame de Candolle. Rather, it was for having es-

3. Quoted in Slobodchikoff (1976, p. 145).

tablished the rules of binomial nomenclature in 1753, whereby each species was named with two Latin words: the name of the genus to which the species belongs, and an adjective that characterizes the species.[4] Linnaeus included two other rungs in his classification—order and class—although he did not consider them to be to natural biological divisions, unlike genus and species. Contrary to received wisdom, his classification was not based on rigid principles. Organs of reproduction were important when it came to classifying plants—the number of stamens to determine the class and the number of pistils for choosing the order—but other criteria could come into play as well, such as their general shape. His classification of animals was based on previous classifications, although he made some important changes—he placed whales in the class of mammals and grouped humans and monkeys together in the "anthropomorphous" order, which he renamed "primates."

His ideas were not set in stone: he corrected errors in his first classifications and progressively came to realize that similarities between species made the classification process a very difficult one. The creation of hybrids, which he undertook himself, also blurred these distinctions. By the end of his life, he had come to think that species were not all created as they are today, and that some had formed through the transformation of other species.

Linnaeus was not an anatomist, but like many of his predecessors he tried to find connections between the anatomy of animals and that of plants, between the skin of animals and the bark of trees. He was a remarkable observer, who was as interested in the customs of the peoples he encountered as in the plants he studied. He was able to quickly identify similarities, which no doubt explains the value of his classifications and the

4. Hoquet, 2005.

success they would eventually achieve. He was the first, before Goethe, to propose that the processes by which leaves and flower petals develop are similar. He explained the mechanism by which pearls are formed in oysters, through the accumulation of mother-of-pearl around an irritating object, long before the Japanese started producing artificial pearls using the same principle.

While I was hesitant to call Theophrastus the father of ecology, Linnaeus is a much better candidate. He ascribed a great deal of importance to the geographic distribution of plants and to the different ways that plants associate with one another. In nature, he noted a balance between plants and animals, predators and prey, which he called the "economy of nature," and which he believed was controlled by the hand of God. He was the first to draw up the biogeochemical cycle of matter through the lens of living things.

I stated that the success of Linnaeus's classification could be explained by his focus on sexual characteristics, comparing stamens and pistils to husbands and wives and the flower to the marriage bed. However, the real reason for his success is that Linnaeus arrived at precisely the right moment. Descriptions of new plants and animals from recently explored continents made the old classifications obsolete. It was in Holland, the port of entry for many of these new species, that Linnaeus came up with and perfected the principles of his classification. The eighteenth century was a century of classification, no doubt in part because of the accumulation of new observations.[5]

Linnaeus's works grew in relation to this deluge of data and his concomitant efforts at classification. There would be 12 editions of *Systema naturae* during his lifetime, and a thirteenth

5. Müller-Wille and Charmantier, 2012.

published posthumously. The first edition was 11 pages long, while the last edition was made up of 10 volumes and totaled 6300 pages.

Linnaeus's interest in classification went beyond its value as merely a source of knowledge. He wanted to provide his country with the means of acclimatizing new plants, whose cultivation would provide greater autonomy and reduce the country's reliance on imports.[6]

Human beings were not excluded from this interest in classification, and Linnaeus was one of the first to group human beings into five races—Africans, Americans, Asians, Europeans, and the "Monstrous" (in which he included Hottentots, Patagonian giants, and the "cretins of the Alps"). In the first editions, he also mentioned two other human species, including troglodytes, which he would later delete owing to a lack of proof of their existence. His efforts to classify the human species would be continued by Johann Friedrich Blumenbach (1752–1840) and Immanuel Kant (1724–1804) in 1775, and would haunt the thoughts of biologists in the nineteenth century. Blumenbach is considered the father of physical anthropology for introducing precise methods for measuring, in particular when it came to the shape of skulls (craniometry). France was the only country where Linnaeus's classification was received with circumspection, under Buffon's influence.[7] Georges-Louis Leclerc de Buffon (1707–1788) first became known as a physicist and mathematician, and he was among those to introduce Newton's ideas in France. His *Théorie de la Terre* (Theory of the Earth), published in 1749 as a part of *Histoire naturelle* (Natural history), extended the age of the Earth to 100,000 years, which led to

6. Koerner, 1999.

7. Roger, 1963, 1989; Schmitt, 2007.

some difficulties with religious censorship. In this original work, Buffon attempts to provide a complete history of the Earth, noting the progressive emergence of new living things, though this does not mean that he had an evolutionary view of life. As we will soon see when we describe his model of reproduction, organic molecules capable of forming organisms were, for him, present everywhere in nature.

In 1739, Buffon was appointed curator of the Jardin du roi, or King's garden, and it is in the context of this work that he came up with the idea of publishing a catalog of the garden's collections. This project would eventually become *Histoire naturelle*, written in collaboration with Louis Daubenton (1716–1799), who was responsible for the anatomic descriptions of animals.

The first volumes published dealt with quadrupeds (12 volumes) and birds (9 volumes). Two editions of the work on birds were made available, one "standard" edition and another enhanced with magnificent color plates, which required that each image be painted by an artist. *Histoire naturelle* had 36 volumes in total.

In his opening remarks, Buffon described the principles that guided the production of *Histoire naturelle*, which he compared to those of Linnaeus.[8] He believed that there was no such thing as a natural classification. In his eyes, all classification was arbitrary because it separated that which in nature exists as a continuum. Thus, in his view, the best classification was one based on common sense, ordering animals based on their similarities and the ways in which they are interesting to humans. Buffon made fun of those who sought a perfect classification by examining the number of pistils or stamens under a microscope, rather than looking at the organism as a whole. In his view, they

8. Buffon, 1986.

were projecting their simple ideas onto nature in a way that did not take into account its richness nor its complexity.

Buffon's *Histoire naturelle* was a huge success. It provided original information both on animals' ways of life and anatomy, was richly illustrated, and Buffon's style of writing was universally lauded. He would create a loyal audience of readers, as large as Linnaeus's, though he took a different approach.

Also in the same vein as Linnaeus, Buffon ventured into anthropological territory in describing human races. He would cast doubt on the stability of species, which he believed could degenerate, and thus came up with the hypothesis that a donkey is a degenerate horse. However, he refused to consider the possibility that monkeys were degenerate humans.

It is tempting to think that these reflections were sowing the seeds of evolutionary thinking—albeit in a somewhat restrained way when it came to humans, out of fear of entering into conflict with the religious authorities. However, there is no reason to doubt Buffon's sincerity when he stated that there is a wide gulf separating the most primitive human beings from animals. Moreover, in Buffon's view, the different races of humans whose characteristics and temperaments he described were the result of reversible adaptations to the specific environments in which they lived. However, his ideas on the transmutation of species were somewhat limited in their scope. As he saw it, it was simply the result of degeneration from the influence of the environment, and he only asked the question of whether monkeys descended from man, not the reverse. Linnaeus, whose belief that species were fixed would be ridiculed in the nineteenth century, went as far as Buffon when he suggested that the original number of species was smaller than the number of species in existence today.

Reproductive Physiology

The excellent work in descriptive anatomy that started in the previous century continued throughout the eighteenth century. One of the people who came to represent this work was Albrecht von Haller (1707–1777), a professor at Göttingen who was also a diplomat for the city of Berne. His anatomic work was universally praised. He dissected more than 400 cadavers and identified many anatomic structures, a number of which bear his name, and compiled many of the anatomic discoveries that came before him. His work will be remembered for his interpretation of irritability (of muscles) and sensitivity (of nerves) rather than the arguments he made with respect to the distinction between these two characteristics that are unique to living things. He did not speculate as to the origins of these characteristics, but used them to describe the functioning of organisms, a view that was in keeping with the moderate vitalism that would predominate at the end of the eighteenth century.

The work of René-Antoine Ferchault de Réaumur (1683–1757), *Mémoire pour servir à l'histoire des insectes* (Memoirs serving as a natural history of insects), published in six quarto volumes, can be seen as a continuation of Swammerdam's work. Though it is not as rich in anatomic descriptions of insects, Réaumur's work was innovative in that it examined their social lives. It is also worth mentioning the monograph of Dutch engraver turned naturalist Pierre Lyonet (1707–1789) on caterpillar anatomy.

The eighteenth century also saw its share of metaphysical conjecture. Julien Offray de La Mettrie (1709–1751) did not conduct any experiments, nor was there anything original in his ideas. Indeed, his mythical account of the formation of organisms was copied from Lucretius's *De rerum natura*. After a sojourn in Holland, he published *L'homme machine* (Machine man), an extension of Descartes's mechanistic view of the living world to human beings, in which he denied the existence of a rational and immortal soul. He thought that monkeys, if trained, could stand up on their hind legs and begin speaking, which Dutch anatomist Petrus Camper (1722–1789) showed to be impossible, given the animals' anatomy. Though his pamphlet gained some renown, his role in the development of biological thought is questionable.

Many biologists from the period focused their work on reproduction. Two discoveries that may seem anecdotal to us today helped focus the debate, but also masked some real developments.

The first was the discovery that some simple organisms such as hydra were able to regenerate, described for the first time by Abraham Trembley (1710–1784) in his monograph on the freshwater polyps, published in Geneva. His observations were reproduced by his compatriot Charles Bonnet (1720–1793) in annelids and by Lazzaro Spallanzani (1729–1799) in his work on the regeneration of limbs in salamanders. Trembley also described the hydra's ability to reproduce through budding. Contrary to what we would believe, the discovery of these regenerative phenomena did not present a challenge to preformation theory. However, it meant germs had to be dispersed throughout the organism and had to be living particles, which La Mettrie and Buffon both discussed. These living particles echoed the seeds of the alchemic tradition, but they are also the successors of Leibniz's monads, or spiritual elemental particles.

The second discovery was that of parthenogenesis in certain insects by Bonnet, whose work is in the same tradition as Swammerdam's. As he was unable to observe things under the microscope when he was very young owing to an eye disease, Bonnet quickly turned to philosophical speculation. Parthenogenesis is reproduction by females only, without fertilization of the eggs by males, and was perfectly in keeping with preformation theory and that of encasement of germ cells. It also justified the rejection of spontaneous generation.

This did not stop Bonnet from thinking that living beings transformed, not through degeneration, but by continually progressing up a "ladder."[9] These transformations were thought to have occurred alongside the catastrophes that had stricken the Earth, of which the last was the flood described in Genesis. In his view, these catastrophes were predetermined and only revealed the ladder of beings.

Bonnet's preformationism squares well with his mechanistic view of the living world and his opposition to vitalism, whose only merit in his view was as a word to fill gaps in explanations. More surprising was his faith in the transformation of organisms—the word "faith" is used intentionally here because these transformations were part of a divine plan, in Bonnet's eyes. Lamarck and Cuvier, who expressed his admiration for Bonnet's work, built on his ideas.

Buffon proposed a model of reproduction that was very different from Bonnet's. In his view, organic molecules—not in the modern sense of the term, but referring to the living particles previously mentioned—were a by-product of organic decomposition. These molecules were found everywhere in nature and were said to be responsible for spontaneous generation of infusoria (small microorganisms). As he saw it, they carried the seeds for reproduction, and embryonic development was reproduction of the organism from these organic molecules. Accord-

9. Anderson, 1976.

ing to his model, two principles guided the formation of the embryo: the affinity of organic molecules for other organic molecules of the same type, and the existence of an interior mold. The first principle comes from Hippocrates, but also from applying Newton's law of attraction to chemistry, which chemists did during the second half of the eighteenth century (Buffon was a staunch Newtonian). The idea of the interior mold is borrowed from naturalist Louis Bourguet (1678–1742). Much as a mold is used by a sculptor to create the shape of a statue, the interior mold did the same for the internal shape of organisms.

Buffon's belief in spontaneous generation, supported by his ideas on organic molecules, found support in the 1743 work of English priest John Turberville Needham (1713–1781).[10] Needham conducted an experiment that would have a bright future in the history of biology. He took a phial containing lamb juices and placed it on a brazier to destroy any microscopic organisms present, whether within the phial or on its surface or the stopper. Despite this heat treatment, the liquid became cloudy and infusoria developed.

Spallanzani redid the same experiment between 1765 and 1776, simply ensuring that the phials were well sealed and heating them for longer. Under these conditions, no infusoria appeared in the flask, which was in keeping with Spallanzani's preformationist ideas. Needham responded by stating that the more brutal treatment that Spallanzani had subjected the phials to must have altered the generative force of the organic molecules or the air in the phial. Spallanzani could not respond to Needham's objections, and it would take the collective experimental imagination of nineteenth-century scholars, including Pasteur, in order to do so.

10. Brock, 1999.

Spallanzani, whose work seemed to reinforce the egg-based preformation theory of which he was an enthusiastic supporter, also conducted experiments that demonstrated the role of sperm in fertilization. By dressing male toads in small underpants, he prevented fertilization and development of the female's eggs. He then collected drops of ejaculate from the underpants, with which he was able to directly fertilize unfertilized eggs. A few years later, he would achieve the same results with mammals: he would fertilize a female dog by introducing semen obtained from a male into her vagina. Spallanzani was a peerless experimenter,[11] as his work on digestion showed. He was able to demonstrate the role played by gastric acid in digestion, as distinguished from any mechanical action of chewing or grinding, which Réaumur failed to do. The experiment involved making animals swallow tubes pierced with holes and containing meat, which he then collected and examined the contents.

The debate between preformationism and epigenesis would be settled in a very different way, with a return to direct observation of embryonic development in chickens.

In 1759, Caspar Friedrich Wolff (1734–1794) published his theory titled *Theoria generationis*, in which he asserted that the structures of the embryo are formed de novo and do not preexist in the egg. A debate with Haller, supported by Bonnet, would continue for several years, each bringing forth new observations.[12] Haller had initially been a supporter of epigenesis, but he later converted to preformationism. The formation of certain embryologic structures—for example, blood vessels, the embryo's different membranes—would be at the heart of this debate.

11. Rostand, 1951.
12. Roe, 1981.

Many of Wolff's observations were correct, but he did make some errors, which Haller, an excellent anatomist, corrected. The difficulty in settling the debate revolved around whether structures *existed* at certain stages of development, while experiments could reveal only whether those structures could be *seen*. Supporters of preformation theory could always state that the structure was there but invisible. Indeed, Wolff was so keenly aware of the problem that he deemed the use of a microscope to be of little value.

Wolff's second work, published in 1768, *De formatione intestinorum* (On the formation of the intestines), helped to move things on from the previous debate.[13] In it, Wolff described the specific mechanisms that led to the formation of the intestine: the appearance of a membrane that folds over on itself to form the beginnings of the digestive tract. Thus, he proved the existence of transitory structures that appear and then disappear over the course of development. With this first draft of specific mechanisms for embryogenesis, which would become germ layer theory at the beginning of the nineteenth century, preformation was no longer plausible. Uncovering a specific mechanism for embryogenesis invalidated any notions of preexistence.

But Wolff's scientific worldview, like that of many other scholars from the period, does not square well with what his decisive role in the emergence of modern embryology might lead us to expect. Though Wolff stated from the beginning that he was determined to fight against preformation theory, it was because he was convinced that embryonic development was due to the action of an "essential force," which he explicitly recognized as having borrowed from Stahl, although he did not identify it as the soul. Like many naturalists before him, he was

13. Wolff, 2003.

also convinced that plants and animals shared a common structure. He sought, with little success, to show that the development of plants and animals followed the same mechanisms.

The final contributions that we will look at are those of Benoît de Maillet (1656–1738) and physicist Pierre-Louis Moreau de Maupertuis (1698–1759). Both of these men seemed to anticipate the discoveries of the following century on the evolution of organisms and hereditary mechanisms. De Maillet, French consul in Egypt, was the author of *Telliamed, une histoire de la Terre* (Telliamed, a history of the Earth). In this work, which began to circulate from 1720 but would be published only after his death in 1748, he envisaged how the mountains progressively emerged from the sea, and how animals also emerged from the sea and transformed into terrestrial organisms. In the following century, Cuvier used *Telliamed* as an example of absurd ideas, not based on serious scientific study, that his rivals Geoffroy Saint-Hilaire and Bory de Saint-Vincent supported.

Maupertuis was a supporter of Newton's theory and took part in an expedition to the Arctic Circle in order to determine the shape of the Earth, a crucial observation that would validate or invalidate the theory. He came up with the principle of least action, although he would later dispute with Leibniz, who also claimed credit for its authorship.

His work in biology is less well known, but François Jacob cited it a number of times in *La logique du vivant* (*The Logic of Life*), calling Maupertuis one of the fathers of genetics.[14] Maupertuis is also recognized as being one of the first to make a connection between the transformation of living species and the process of embryologic development and epigenesis, thus anticipating the work in evolution and development at the end

14. Jacob, 1993.

of the nineteenth century. It would surely have pleased many physicists to see one of their own play a decisive role in advancing biological knowledge.

Maupertuis's ideas were presented in two works: *Dissertation physique à l'occasion du nègre blanc* (Physical dissertation on the occasion of the white negro), published in Leiden in 1745, and *Système de la nature* (System of nature), published in 1754.[15] In these works, he explained the transmission of traits from parents to children, such as the phenomenon of six-fingered hands, which he studied. He used a Hippocratic model no doubt borrowed from Buffon, according to which organic molecules that are transmitted assemble over the course of development based on their desires, attractions and aversions. He used the example of a chemical reaction that is able to create a shape analogous to a tree (known as Diane's tree) and he envisaged that similar reactions could occur over the course of embryogenesis to give shape to the organs. He believed that reorganizing organic molecules could create new organisms, and thus it was possible to envisage how organisms had transformed over the course of the Earth's history.

It is tempting to refer to current knowledge in interpreting de Maillet's and Maupertuis's writings, or to see Maupertuis as the father of theories of self-organization, but a few observations are in order. However original their writings may have been, they did not generate any immediate scientific interest, nor did they form the basis for any schools of thought. Maupertuis's idea of endowing organic molecules with desires, which was a transformation of Leibniz's ideas on monads, would be criticized by Diderot, who preferred to explain the origins of the traits that characterized life as resulting from the complex

15. Maupertuis, 1980, 2001; Sandler, 1983.

arrangement of parts. And for those who have read Lucretius's *On the Nature of Things*, the originality of de Maillet's and Maupertuis's ideas on the origins and evolution of living things is less apparent.

Maupertuis had nothing to say about a process to allow for differential survival of modified organisms. François Jacob places his originality in his idea that new organisms result when their parts assemble in a different configuration from that in the organisms that conceived them. In modern parlance, we would say that there was a different recombination of these parts to create a new whole, or, as Jacob described it, "tinkering."[16] However, from an atomist point of view, where matter is eternal and made up of different atoms, is it not natural to regard newness as resulting from a new arrangement of atoms? Indeed, this idea could already be found in Lucretius's writings.

The Role of Breathing Becomes Clear

While certain functions of organisms, like digestion, were understood at least partially quite early on, others, like breathing, remained a complete mystery for centuries. Air was seen as a cooling agent, but also as the source of animal spirits, whose existence was not called into question from Aristotle until the time of Descartes.

At the end of the eighteenth century, a totally new view would take hold over the course of a few years. Breathing was said to bring the oxygen needed for the combustion of food, which would produce heat for the organism.

The process of analyzing air was begun by Van Helmont, who had introduced the idea of gases and described what he

16. Jacob, 1977.

called sylvan, or forest gas—carbon dioxide. English physician Joseph Priestley (1733–1804) described oxygen in 1776. In keeping with phlogiston theory, he called this new gas "dephlogisticated air." He demonstrated that animals absorbed oxygen and released carbon dioxide, and that a candle would no longer burn in air "diminished" by animals, but that this air could be restored by plants.

Antoine-Laurent de Lavoisier (1743–1794) did not discover any gases himself, but he contributed to explaining and quantifying the phenomenon of breathing.[17] He showed that oxygen represents roughly a fifth of the air's volume. In 1777, he confirmed that oxygen is necessary for animals to survive and that breathing released carbon dioxide, like combustion.

He came up with two hypotheses to explain this: that oxygen was directly converted into carbon dioxide in the lungs or that the gases were exchanged, opting for the second explanation. He interpreted the fact that blood changes color after passing through the lungs as being due to the attachment of oxygen. In 1782 and 1783, he published a long essay on heat, which provided a definition of heat and a means for measuring it through a new experimental apparatus, known as the "calorimeter." The amount of heat released was estimated by the quantity of ice that it was able to melt. With this device, the quantity of heat produced by a guinea pig's respiration was determined, and compared to the quantity of heat produced through the combustion of coal. The values were identical for the same quantity of carbon dioxide produced, and therefore respiration was seen as simply the combustion of food due to oxygen in the air.

Lavoisier's work, like that of Priestley, was not limited to the study of respiration. Often considered to have instigated the

17. Bensaude-Vincent, 1993.

chemical revolution, Lavoisier made changes to chemical nomenclature. He showed that water is formed by bringing together oxygen and hydrogen, the latter having recently been discovered by English chemist Henry Cavendish (1731–1810). Lavoisier put an end to phlogiston chemistry, but introduced another concept that would also be done away with later—caloric, meaning that which allowed substances to change into a gaseous state. Lavoisier's work was a product of its time. The bringing together of respiration and combustion occurred at the moment when the Industrial Revolution was taking off and consumption of wood and then coal for steam engines was rising. It is difficult to see this as only a temporal coincidence.

Lavoisier's life was firmly rooted in his period. He was appointed farmer-general in 1768 and commissioner at the Régie des poudres et des salpêtres in 1755. He played an important role in modernizing agriculture and improving hygiene in prisons and hospitals, in particular through his work within the French Academy of Sciences. He saw no value in mesmerism, a form of therapy based on animal magnetism, invented by the German physician Franz-Anton Mesmer (1734–1815). He was guillotined on 8 May 1794, along with 27 other farmers-general. Those who wished that his be life spared because of his work were told that the Republic did not need scientists. As well as for his involvement in the ancien régime and the profits he earned through it, Lavoisier also paid the price for his self-assured view of science, which looked with disdain on popular pseudosciences like mesmerism, which certain revolutionaries, like Jean-Paul Marat (1743–1793), adhered to.[18]

Priestley's observations on plants were completed by Dutch botanist Jan Ingenhousz (1730–1799), who showed that oxygen

18. Darnton, 1986.

is produced by the leaves under the action of light. Without light, leaves produce carbon dioxide, as do all other parts of the plant, regardless of whether they are lit. Nicolas Théodore de Saussure (1767–1845), a geologist and also a politician in Geneva, described the precise quantitative relationships between the absorption of carbon dioxide by plants and the production of oxygen.

Historical Overview

Variations on Vitalism

Vitalism did not originate with Stahl.[19] In fact, Aristotle and Galen were vitalists, since they believed that organisms were carriers of immaterial principles. For Aristotle, these immaterial principles explained animal movement and were also the final cause that guided embryonic development. Stahl did not renounce these functions, but he added another one to them, which was to allow the organism to fight against the corruption that perpetually threatened it. Bichat would use this new function a century later when he described life as the collection of functions that resist death.[20] As we will see later, at the beginning of the twentieth century embryologist Hans Driesch would return to Aristotle's conception and would make the embryo's capacity to "aim for" the adult form the foundation of his neovitalist ideas.

The vital principle can therefore fulfill at least three roles: a teleological one, as in the preceding example; causing movement; and ensuring the organism's survival. As we have seen, the vital principle has sometimes been simply a way of explaining an

19. Canguilhem, 1965a; Nouvel, 2011; Normandin and Wolfe, 2013.
20. Bichat, 1977.

organism's traits when they cannot by current understanding be reduced to physical and chemical mechanisms. The only common thread among all of these definitions of the term "vitalism" is the difference that exists between the inanimate world and the world of organisms. It is in this context that we should view its resurgence at the beginning of the eighteenth century. Many naturalists went against Descartes's work and that of the iatrophysicists to reaffirm the originality of the phenomena that characterized the living world. Thus, we can separate positive vitalism, which affirmed the existence of a vital principle and attempted to ascribe these characteristics to it, and negative vitalism, which sought simply to show the inadequacy of physical and chemical explanations of phenomena from the living world.

Biologists in the second half of the twentieth century considered vitalism to be an aberration, as it called on nonphysical entities to explain the characteristics of living things, foregoing a natural explanation. As we have seen, this view of vitalism does not reflect the historical reality that its different forms played. Nor, as Georges Canguilhem has shown, does this view recognize the positive role that vitalism played in the eighteenth century in the progressive building of the science of life that at the beginning of the nineteenth century would come to be known as "biology."[21] Nor does it explain why vitalism was largely accepted by the scholars of the Enlightenment and the authors of the French *Encyclopédie*. It was deemed by some of them to be compatible with their materialist ideas and was firmly entrenched in the thinking of all biologists in the nineteenth century, even those such as Claude Bernard who apparently crossed swords with vitalism. Of course, all this does not mean that vitalism still has a role to play in contemporary biology.

21. McLaughlin, 2002.

Classification versus Evolution

In the second half of the nineteenth century, adherents of the new theory of evolution often portrayed the classifiers of the eighteenth century, Linnaeus foremost among them, as believing that species were fixed. As we have seen, this is a distortion of Linnaeus's ideas, who acknowledged limited transmutation of species.

However, more seriously, this view misses the point that coming up with a rigorous classification was a necessary condition to be able to conceive of the idea that living species could transform themselves. The theory of evolution had no place in a world where the existence of monstrous hybrids was accepted or where the most fantastical transformations were conceivable, as described by Lucretius and Maupertuis. Nor does evolution in the modern sense of the term—as creating new forms—have a place in a "continuous" world where all conceivable living forms are supposed to exist. The development of rigorous classification did not hinder the rise of an evolutionary worldview, but it made it possible. It is because living things are not in a permanent state of transformation—their traits are sufficiently stable to be described—that an evolutionary history of the living world is conceivable, in the same way that the history of human societies came about only when those societies structured themselves.

Classifying Humans

Given the preoccupation with classification amongst eighteenth-century scholars, Linnaeus and Buffon, but also Kant and Blumenbach, came up with lists of the different human races and detailed their physical and behavioral characteristics.[22]

22. For Kant's ideas, see Lagier (2004).

Anthropology developed in the nineteenth century, and a hierarchy of human races would be introduced. This hierarchy was not completely absent from the first classifications, but it was kept in check by two facts that no biologist from the period called into question. The first was the uniqueness of the human species—attempts by Linnaeus to add two other species of human beings would come up against the fact that they just didn't exist. The second was that the influence of the climate and the environment and heredity of acquired characteristics could explain the differences between races, which it was believed would eventually disappear as human beings moved around.

Nonetheless, cracks started to appear in this fragile defense against biological racism at the end of the eighteenth century. There was no evidence that that Black people from Africa transported to the Americas changed their physical characteristics, and likewise, Europeans did not take on the traits of Indians. Descriptions of castes on the South American continent reaffirmed this racial stability, and were also an important step in building the science of heredity.[23] On the other hand, the hypothesis of polygenism, which stated that there was not one but many human species created separately with distinct characteristics, appeared in the second half of the eighteenth century in works from writers and philosophers like Voltaire, before gaining support among biologists like Cuvier.

Priestley and Lavoisier: Only the First Step

Likening respiration to combustion was only the first step toward a fuller understanding of this phenomenon. Lavoisier, for instance, took the comparison only as far as the production

23. Müller-Wille and Rheinberger, 2007.

of heat. There was still a crucial concept missing—that of energy, which would come into play in the nineteenth century with the rise of thermodynamics. Respiration created energy through the combustion of food, and the energy was used by the organism to produce heat, but also, more importantly, to ensure all of its basic functions: nerve activity, muscle contraction, synthesis of all of the components of life, and so on. These gaps in Lavoisier's interpretation, because of the lack of a concept of energy, makes it all the more surprising that he claimed to have measured equal amounts of heat produced for the same quantity of carbon dioxide released in both the combustion of coal and animal respiration.

Contemporary Relevance

A Natural Classification?

The existence of a natural classification of animals and plants was at the heart of the debate between Buffon and Linnaeus. Biologists today, after Darwin, would say that only a classification based on descent can be natural. In essence, Linnaeus and Buffon were both wrong—the former for having thought he had developed a natural classification, and the latter for denying the interest of looking for a better classification. Today, most classifications are based on phylogenies developed by comparing sequences of nucleic acids.

We will put the practical difficulties that we run into in building these phylogenies to one side, to ask a fundamental question: In what way is classification based on phylogeny more "natural" than classification based on the relative utility of plants and animals for humans, as was the practice up until Buffon? If you ask this question to modern taxonomists, you would be surprised at the difficulty they have in providing you with an answer!

This is because Buffon was right: there is no natural classification, because the act of classifying is a purely human activity and is always carried out with a specific goal in mind. What makes classifications based on descent valuable is that, being based on a measurement of "genetic distance," they are the most likely to reflect the overall differences between species. There is merit in this approach, but it does not exclude the fact that for specific uses, other classifications might prove more useful—as in microbiology, where the classifications used today are only partially based on comparisons of nucleic acid sequences.

Comparing Plants and Animals

We have described the repeated efforts of naturalists to find corresponding anatomic structures and physiological processes in plants and animals—for example, skin and bark—or, in Wolff's case, to uncover similarities between embryonic development in animals and that in plants. This persistence can no doubt be explained both by the long tradition of looking for things that "correspond" in nature, which was a common practice among alchemists, and by a tradition in science that attempts to understand the unknown through the known—that understanding the movement of blood, for example, should "naturally" help us understand the movement of sap.

We can poke fun at these naïve attempts, but it is more interesting to delve into the difficulties faced by contemporary biology when it comes to recognizing the similarities and differences between animals and plants. Along with the paralyzing effect that the memory of past errors can often have, the first reason for this is no doubt the almost complete separation of teaching and research in animal biology and plant biology. A work on developmental biology, for example, may contain not a single chapter on plant development; likewise, a work on cell signaling—the study of the different signals exchanged between the compartments of a cell or between different cells—might be limited to examples taken from only the animal world. Moreover, there is a second more profound difficulty—biologists don't know what to do with these comparisons. The reason for this is simple. Since Darwin, comparative anatomy and physiology have been based entirely on phylogenetic relationships. However, the mechanisms involved in signaling and developmental biology were created independently, after the separation between the ancestors of plants and those of animals. Contemporary biologists are therefore as powerless to make these comparisons as naturalists were in the eighteenth century. Indeed, they are in a worse situation, as they cannot refer to a single plan for nature nor the actions of a Creator to explain these similarities.

Maupertuis, the Father of Self-Organization?

We have seen how Maupertuis explained the morphogenesis of embryos by comparing them to the shapes that chemical reactions can spontaneously create. In doing so, he opened up a line of thought that many would follow after him—to explain living phenomena through analogy with processes that we would describe as self-organizing today.

The problem with this type of explanation is that it too often rests on resemblance. The resemblance may be reinforced using mathematical equations—but it remains merely resemblance, so long as the underlying mechanisms are not known. We have seen how the important role given to explanations by analogy in ancient Greece slowed the development of rigorous experimentation. Should supporters of self-organization therefore be happy to have Maupertuis as a forerunner? Or should they be alarmed at how antiquated and weak these types of explanation continue to be?

6

The Nineteenth Century (Part I)

EMBRYOLOGY, CELL BIOLOGY, MICROBIOLOGY, AND PHYSIOLOGY

GIVEN THE DIVERSIFICATION of biological work in the nineteenth century, I have decided to devote separate chapters to functional biology and evolutionary biology.[1] We will examine the history of embryology, cell biology, microbiology, and physiology in this chapter, and we will tackle evolutionary theory, the science of heredity, and ecology in chapter 7.

The key development in this chapter is the emergence of cell theory, though its relationships with the other disciplines are quite different. Cell theory allowed embryologic models to be reinterpreted, without completely transforming them. It was an essential condition for the development of microbiology, though microbiology didn't benefit from the vast quantity of work carried out on the structure of cells. Cell theory did influence physiology in some ways, but it ran into difficulties conforming to the chemical approach to living phenomena that prevailed in the discipline at the time.

1. Mayr, 1961.

The Facts

Embryology Becomes an Established Discipline

The work that was being carried out at the beginning of the nineteenth century can be situated in time after Wolff's work.[2] Christian Pander (1794–1865), who was born in Riga in Latvia, began his studies at the University of Dorpat (today, Tartu in Estonia), and then continued in Berlin, Göttingen, and Würzburg, where, in 1817, he carried out the work on embryonic development in chickens that would make him famous, in collaboration with Baer. He devoted the rest of his life to exploring the world, working on geologic projects, and developing ideas on the transmutation of species supported by his paleontological observations (a sort of Lamarckism that included the notion of metamorphosis from the German *Naturphilosophie*).

He confirmed Wolff's observations and developed germ layer (embryonic) theory, based on the foundations laid out by Wolff. Each of these layers, which first appears as a simple membrane, forms progressively over the course of embryonic development, and through different processes, such as invagination and folding, gives rise to the different structures in the embryo. The endoderm, one of the three germ layers, creates the digestive tract and the stomach, but also the pancreas and the lungs. Germ layer theory is both a mechanistic explanation of development and a tool for comparing embryonic development in different organisms.

Karl von Baer (1792–1876) was born into a German family who settled in Estonia. He also studied in Dorpat. After collaborating with Pander he continued his work in Königsberg, then in Saint Petersburg (the Baltic countries were part of Russia at

2. Churchill, 1991; Gilbert, 1991; Dupont and Schmitt, 2004; Schmitt, 2006.

the time). He would also take part in explorations in Russia, where he would become a geographer, anthropologist, and ethnologist. In 1827, he described mammalian eggs (in dogs), going further than De Graaf, who thought he had found the eggs when in fact they were the follicles. The comparisons he made between the embryonic development of different species showed that the more different two species were, the earlier you have to go back in their embryonic development to find similarities. From this he concluded that over the course of embryogenesis more general characteristics form earlier (for example, if the organism being studied is a mammal, the characteristics that are shared by all mammals), before characteristics that are specific to the species being studied. Embryogenesis goes from the general to the specific, from homogenous to heterogenous, an idea that is the total opposite of preformationism. Baer adopted Cuvier's idea that there are several levels of organization (types) in animals (see below), and thus comparisons must be made within these types. Baer did not have any evolutionary ideas. Indeed, by the end of his life, he had become a staunch opponent of the theory of evolution proposed by Darwin.

Baer's model of development contrasted with the model proposed by Johan Friedrich Meckel (1781–1833), working at the University of Halle in Germany, and by Étienne Serres (1786–1868), who held the comparative anatomy chair at the Muséum d'histoire naturelle in Paris. Meckel and Serres were both influenced by Lamarck's transformist model. They believed that organisms progressed along the great chain of being, both in their successive transformations over the course of generations, as Lamarck hypothesized, and during embryonic development. An organism was essentially the sum of the different stages its ancestors had gone through. This theory would be taken up and expanded by Ernst Haeckel following the publication of Darwin's evolutionary model. There are some common features between Baer's model and that of Meckel and Serres—such as, for example, the similarity of all embryos at the beginning of development. However, there are also differences. Meckel and Serres thought the embryo of a species located higher up the great chain of being could resemble the adult organism of a simpler species, while Baer thought that only the embryos from these two species would be similar.

Two other contributions to embryology are worth mentioning. The first is that of Martin Heinrich Rathke (1793–1860), Baer's successor at Königsberg. He discovered that mammalian embryos form transient branchial slits and arches (still called "visceral arches") that resemble a fish's gills, an observation that would later be expanded by Karl Reichert (1811–1883). Similarly, he showed that over the course of embryonic development, the kidneys replaced another transient organ that performed the same function—the pronephros. These two discoveries were important partly because they put the final nail in the coffin of preformation theory by showing that embryonic development is a complex process by which an organism is formed, with transient structures, and not the revelation of a preestablished plan. The discovery of branchial arches seemed to be in full agreement with Meckel and Serres's theory, a magnificent example for Haeckel's recapitulation theory, and an (apparent) confirmation of Benoît de Maillet's (and others') imaginative powers.

The final, perhaps most important, and certainly most unrecognized player is Robert Remak (1815–1865). A student of physiologist Johannes Müller, he worked in Berlin with Rudolf Virchow (1821–1902), but on account of his Polish-Jewish roots he would never receive a position at the university. As we will see, it was Remak and not Virchow who first came up with the idea that all cells come from the division by fission of another cell. His contribution to embryology is to have reconstructed embryonic germ layer theory, of which he reduced the number of layers from four to three, by incorporating results from cell theory.

The Emergence of Cell Theory

Cell theory did not emerge out of a conceptual vacuum.[3] In the late eighteenth and early nineteenth centuries, many biologists attempted to define a structural component that might be common to plants and animals. Cell theory was novel, not in stating that cells existed, but in making them the sole constituent parts of living organisms.

Before cells, fibers seemed likely candidates. The presence of fibers in organs was first noted by Francis Glisson (1599–1677), a professor of physiology at Cambridge. Descartes believed that what he called "threads" played an important anatomic and physiological role in organisms, and Baglivi also noted their presence. However, it was Haller, a physiologist, who saw them as general structural components, stating that "the fiber is for the physiologist what the straight line is for the geometrician."[4] He distinguished three types of fibers: fibers that make up muscle tissue, nerves, and what today we would call connective tissue. The physicist and chemist from Florence Felice Fontana (1730–1805) thought he had discovered this basic component of organisms in what he described (in 1765) as tortuous primitive cylinders.[5]

Between these early attempts and the gradual emergence of cell theory, there is the work of Xavier Bichat (1771–1802), and

3. Harris, 1999; Duchesneau, 2000.

4. Quoted in Harris (1999, p. 18).

5. Harris, 1999, p. 21.

his discovery of 21 tissues that, when combined, were said to form all known organs. Bichat was not the first naturalist to refer to tissues, but he was the first to accurately describe them.

Despite the recognition that Bichat's work received, no doubt due to its literary qualities and the considerable influence that it had on French biology, it is a strange work—noticeably out of step with the direction that biology would later take. Bichat's first work, published in 1799, was titled *Traité des membranes en général, et des diverses membranes en particulier* (General treatise on membranes and various membranes in particular). It showed that tissues are first and foremost membranes, which can be revealed by the anatomist's scalpel. Bichat's objective was to reveal the principles of organization that explained the origins and characteristics of diseases. Not only are tissues (membranes) affected in different ways by diseases, but there is also what Bichat called "sympathy" in tissues, which explains why a painful sensation can be felt in distant parts of the same tissue, and what we would call "regulation"—activation of one part of a tissue can reduce activity in another part. Tissues are the essential components of an organism. Bichat was not interested in the details of their structure—he rejected the use of microscopes, because, in his view, they show us what we want to see. Indeed, his books contain no illustrations. Only functions—and especially dysfunctions—were of interest to him.

The philosopher of science Georges Canguilhem indirectly and involuntarily explained the specific role this work played. In several of his books, Canguilhem stated that the study of diseases was at the origin of all research in biology. I believe this general assertion to be false—many works in biology have not used diseases as a starting point—but it applies perfectly to Bichat's work. His ideas were derived from the more than 600 dissections he carried out on individuals who had suffered from various illnesses. He described structures and the ways in which they changed as they were progressively discovered by anatomists working their way toward the innermost parts of the body. Bichat's work was not on the road that would lead to our modern notions of living beings. He was the heir of Barthez and of the Montpellier school of vitalism. Auguste Comte (1798–1857) would make tissues the basic level of organization in living things, and, like Bichat, he rejected the use of microscopes and disagreed with cell theory. Both Comte and positivism hindered the development of cell theory in France. But Bichat's work would stand the test of time, and it remains a valuable contribution in the world of medicine.

We have noted the first observations of cells in the seventeenth century. And descriptions of globules and vesicles would continue unabated. Sometimes these were related to cells, but more often they were artifacts or supracellular structures. This was probably the case with the vesicles that Wolff observed in the embryologic development of both animals and plants.

To adopt cell theory was not simply to see cells (and only cells) inside an organism. It required making those cells the foundational elements of an organism, elements from which it was built. For some biologists, cells were only the fruit of a splitting of a preexisting, continuous membrane in the organism.

The emergence of the first cell theory occurred in three different locations—in France, then in Breslau (modern-day Wroclaw, then a German city, today located in Polish Silesia) with Jan Evangelista Purkinje (1787–1869), and in Berlin with Theodor Schwann (1810–1882) and Matthias Schleiden (1804–1881). The arrival of a new generation of microscopes that corrected chromatic aberrations (i.e., the formation of distinct images for light rays of different wavelengths) partly—but only partly—explains the rapid succession of work in this area.

The first author mentioned in the genesis of cell theory is the French physiologist Henri Milne-Edwards (1800–1885). In 1823, he observed identically sized globules both in animals—in muscles and in the brain's white and gray matter—and in plants. He did not use anything to fix or stain the samples, however, which casts some doubt on the nature of his observations.

Henri Dutrochet (1776–1847), imitating Milne-Edwards, also described globules in plants and animals, though he differed with the latter as to their size. Dutrochet did not mention the presence of a nucleus, and believed that the globules reproduced through the formation of new globules inside preexisting ones. He succeeded, by treating plant tissue with concentrated nitric acid and boiling water, in separating the cells. This led him in 1837 to assert that the concept that turned cells into the foundational components of organisms was henceforth established.[6] He was particularly interested in phenomena he identified that were related to osmosis, which were linked to selective permeability of biological membranes.

The contribution of François-Vincent Raspail (1794–1878) is not well known in the international literature. However, according to his admirers, he came up with the notion of the microbial origin of disease—he spoke of "animated" pathologies—and of the cellular organization of living things. He was first and foremost a chemist, and developed methods for identifying the chemical composition of animal and plant tissues, such as staining starch and glycogen with iodine. He described cells as the centers of chemical activity in organisms. In 1825, he was the first to use the famous phrase "omnis cellula e cellula" (all cells come from a cell),[7] which Virchow would later borrow. However,

6. Vallade, 2008.

7. Quoted in Harris (1999, p. 33).

Raspail had envisaged a sprouting of cells out of existing ones, and not cellular division. Furthermore, his contributions were in part overshadowed by his many other activities, both in the scientific and medical spheres and in the political arena. He was an effective proponent of hygiene theory and of what we would call antiseptics—we owe the widespread use of camphor to his work. A staunch and active French republican, he spent many years in prison under successive regimes in France during that period.

Pierre Turpin (1775–1840) also subscribed to the idea that living organisms are formed of globules, which he thought were filled with a substance called "globuline." Like Dutrochet, he believed that new globules formed inside existing ones. He was convinced that as soon as nature had become organized, it was "globulized." One of his articles includes an epigraph by Leibniz, which helps make the connection between his concept of the important role of globules and Leibniz's monads.

Did the lack of any references by German biologists to these initial observations prevent, as the Quebecois historian and philosopher of science François Duchesneau believes, the recognition of the early influence of French researchers in the genesis of cell theory? We will answer this question below, but, for the time being, let's simply note that there was no particular interest among French biologists in the internal structure of cells, while descriptions of the nucleus and the nucleolus and questions surrounding their functions played an important role in the genesis of cell theory in Germany.

The history here is not simple, either, since those of the Breslau school claimed that their observations had come before those of Schwann and Schleiden, who are generally credited with being at the origin of cell theory.

After having worked in Berlin with Johannes Müller, Purkinje founded a school of microscopic anatomy in Breslau. Purkinje is known for having discovered many anatomic and microanatomic structures, including the large cells of the cerebellum that bear his name. He was the first to describe the germinal disc and the nucleus of the egg in chickens—the cell nucleus in plant cells had been described in 1833 by Robert Brown (1773–1858) and the results published in the Linnaean Society journal.

Purkinje and his students noted the presence of corpuscles in many tissues, including epithelia, bone, and cartilage. His collaborator Gabriel Gustav Valentin (1810–1883) was the first to describe the presence of nuclei and nucleoli in animal cells; the nucleolus had already been observed by Rudolf Wagner (1805–1864) in plant cells in 1835. In 1832, the Belgian biologist Barthélemy Charles Du Mortier (1797–1878) had proposed that cells divide by splitting. Valentin took up this idea, though he did not rule out the possibility that cells were formed endogenously. Valentin was also the first to propose cell theory. In an article that would be awarded a prize by the French Academy of Sciences, he described the presence of cells (and their nuclei) in many tissues of an organism.

However, history has decided that Schwann was the one who unveiled the theory in a work published in 1839, the second part of which bears the evocative title "On Cells as the Basis of All Tissues of the Animal Body." He drew from observations made by his professor, Johannes Müller, in 1835 on the particularly obvious cellular organization of the notochord. The theory was enhanced by similar results obtained by Schleiden in observing plants: cell theory, as Schwann himself called it, applies to both plants and animals.

However, the cell theory proposed by Schwann and Schleiden is markedly different from current cell theory. According to their theory, the nucleolus generates the nucleus (or cytoblast), which then generates the cell. They thought the formation of new cells occurred not through splitting, but inside preexisting cells in plants and in the extracellular fluid in animals. In both cases, cells were said to come from the "cytoblastema"—the name given by Malpighi to living matter—present both inside and outside cells. Schwann compared the formation of nucleoli, nuclei, and cells from the cytoblastema to a "crystallization," an analogy that would be heavily criticized.

From 1841 onward, Remak stated that the primary means of genesis of cells was binary fission. But it would not be until the end of the 1850s that Virchow would put aside his misgivings regarding Remak's hypothesis, reintroduce Raspail's phrase "every cell is derived from a cell," and give cell theory the form and the prominence that it enjoys today.

Is it unfair to credit Schwann and Schleiden with the development of cell theory at the expense of Purkinje and Valentin? Purkinje's Czech nationalism did him a disservice, as did Valentin's Polish-Jewish origins. Johannes Müller, who headed the Berlin group, was able to promote Schwann and Schleiden's work, whereas the Breslau group shared its results only during conferences or in theses. Moreover, Purkinje did not seem convinced that cells were the only basic elements of life—alongside cells, he also mentioned fibers and fluids. Furthermore, Schleiden and Schwann attributed an essential role to the nucleus (particularly in the formation of cells), which Purkinje did not.

Cell theory was not just a reinterpretation of previous observations. It very quickly proved its usefulness in describing and explaining a disease known since antiquity—cancer.[8] Once the idea that cells could form directly from the extracellular cytoblastema was abandoned for good, Müller, Remak, and Virchow reinterpreted cancer as *cellular pathology*, to cite Virchow's

8. Rather, 1978.

eponymous work of 1858. Even if he were wrong in believing that cancer always originated in connective tissue, Virchow contributed to establishing a program of research that would prove particularly fruitful. Leukemia and lymphoma, which remained poorly understood, were shown to be specific forms of cancer. Benign tumors and other lesions were differentiated from malignant tumors, and the different stages in the development of a cancer cell were gradually described. In 1840, Jakob Henle (1809–1885) described the process of the spread of cancer cells from a tumor toward other parts of an organism, known as metastasis. Wilhelm von Waldeyer (1836–1921) distinguished carcinomas of epithelial origin from sarcomas; they were proven to be responsible for the vast majority (80%) of cancers in humans. In 1875, Julius Cohnheim (1839–1884) proposed a theory that differed from Virchow's to explain the origins of cancer. In his view, it resulted from cells left behind during embryonic development that undergo abnormal development in the adult organism—an idea that has gained popularity again at the beginning of the twenty-first century.

Within a few years, all of the necessary knowledge had been acquired to determine the tissue from which a tumor originated and its stage of development by examining tumor biopsies, and thus to provide a prognosis on the development of the disease. However, this did not mean that these tests were immediately put in place in hospitals.

After the improvements in descriptions of cells of the 1830s, a second wave of discoveries was made beginning in the mid-1870s, thanks to progress in the preparation of biological material for observation—namely, creating thin sections of tissues and new methods for fixing and staining.

In 1875, Eduard Strasburger (1844–1912) described the evolution of the nucleus over the course of plant cell division. He showed that the nucleus itself disappears, but that the nuclear material remains and plays a role in the formation of two new nuclei. Walther Flemming (1843–1905) confirmed this result in amphibian larval cells, and introduced the term "chromatin" to describe the nuclear substance that remains despite the transient disappearance of the nucleus. He observed that this chromatin is briefly arranged in the shape of rods, which Waldeyer named "chromosomes" in 1888. Theodor Boveri (1862–1915) described the structure of centrosomes, the orchestral conductors of cell division.

Simultaneously, and in the same competitive spirit, the mechanism for fertilization was (finally) described. The initial observations had been made in plants—brown algae—in 1854 by Gustave Thuret (1817–1875), and were added to by the work of Nathanael Pringsheim (1823–1894) and Anton de Bary (1831–1888), who demonstrated that fertilization was the result of the fusing of two cells.[9] In 1875, Oscar Hertwig (1849–1922) demonstrated that there were two nuclei present in sea urchin eggs at the very beginning of embryonic development. He believed that one of the nuclei—the one located on the periphery of the egg—came from the spermatozoon, despite not having been able to directly observe the sperm penetrating the egg. The Swiss zoologist Hermann Fol (1845–1892) confirmed Hertwig's results in starfish. Four years after Hertwig's first observations, Flemming was able to describe the entry of the spermatozoon's nucleus into the egg. Using a parasitic nematode worm from horse intestines, which has particularly large cells, between 1883 and 1887 the Belgian scientist Édouard van Beneden (1845–1910) demonstrated that the number of chromosomes remains constant in all of an organism's cells, with the exception of

9. Vallade, 2008.

reproductive cells, in which it is reduced by a factor of two. This observation was confirmed by Boveri in 1887. This reduction in the number of chromosomes was the result of a process of reductional division called "meiosis." Its existence was also demonstrated in plants by Eduard Strasburger in 1884. The importance of this phenomenon in the recombination of inherited characteristics over generations would be fully understood only thanks to the observations of the geneticists William Bateson and Thomas Hunt Morgan at the beginning of the twentieth century.

The fusion of the two nuclei after fertilization described earlier can be interpreted as reinstating the usual number of chromosomes. In 1885, Carl Rabl (1853–1917) demonstrated that chromosomes remain separate entities, even though they become invisible during certain stages of cell division. Through a series of experiments carried out on sea urchins beginning in 1889, Boveri showed that the presence of all chromosomes was necessary to the embryo's development.[10]

Though the use of microscopes shed light on the structure of the nucleus, this was not the case for the rest of the cell. The existence of a membrane separating cells from their environment remained unknown until the end of the nineteenth century. The cytoblastema—the matter from which living cells are formed—was renamed "protoplasm" (or "sarcode," by Félix Dujardin). Its nature remained a mystery, but it was said to be responsible for the specific chemical transformations that make up life. Flemming believed it was made of fibers—a strange shift in thinking from seeing fibers as the fundamental components of living things to their being a level below cells. Others, such as Albert von Kölliker (1817–1905), believed that protoplasm

10. Laubichler and Davidson, 2008.

was formed of granules called mitochondria, which were later proven to exist, though their function turned out to be quite different. The same was true of the Golgi apparatus, which was discovered by the Italian biologist after whom it was named.[11] Our rich imaginations often deceptively compensate for the limits of optical microscopy.

German biologists were dominant in this succession of discoveries. This can be attributed to the spirit of competition that dominated the various German universities, and the support of the powerful German chemical industry, which produced dyes for experiments. We can of course also add to this the unfortunate legacy of Bichat, prolonged by the philosopher Auguste Comte as well as the biologist Claude Bernard, which consisted of a mistrust of images from microscopes and the erroneous interpretations that they could lead to. There was also an antipathy toward the "general ideas" whose investigation was advocated by German *Naturphilosophie*, and from which cell theory seems to have arisen.

A new, remote scientific space—the marine station—would become increasingly important in biological research. It could provide researchers with year-round access to organisms whose specific characteristics—cell sizes, transparency, and the ease with which their fertilization could be observed—were well suited to the research being carried out. The zoological station in Naples, established in 1872 by the German biologist Anton Dohrn (1840–1909), is where Theodor Boveri would carry out his aforementioned work. Lab benches at the station were rented out to universities that requested them. Researchers had access to the equipment and chemical reagents necessary for their research, as well as a well-stocked library. Other stations

11. Dröscher, 1998.

were established in France, Germany, and England. The golden age for these stations would come at the end of the nineteenth century with the development of experimental embryology.

The Rise of Germ Theory

Starting in the mid-1870s, in a period spanning several decades, scientists identified the "microbes" believed to be responsible for the most serious infectious diseases in humans.[12]

Infectious diseases had been known since antiquity and measures were progressively taken to avoid contagion, such as the isolation of lepers and quarantine measures for boats coming from regions where epidemics were rife. However, we should not assume that the notion of "pathogenic germs" emerged spontaneously. Indeed, environments such as swamps were also considered to be sources of miasmas and diseases. The hypotheses of Girolamo Fracastoro and Raspail that infectious diseases were caused by germs remained speculation. The term "germ" was a vague one; it seemed to exude more a whiff of alchemy than a clear notion of pathogenic organisms.

More significant was the late eighteenth-century demonstration by two Italian naturalists, Felice Fontana and Giovanni Targioni Tozzetti (1712–1783), that some plant diseases are caused by parasitic fungi. These two authors deserve credit not for having shown that fungi can live as parasites on plants, which was already known, but that plant diseases can be caused by parasites. Curiously, this observation does not seem to have led to the idea that the same type of mechanism might explain certain human diseases. While the similarities between animals and plants were a repeated trope in anatomy, physiology, and

12. Brock, 1999; Berche, 2007.

embryology, nothing of the sort occurred in this case. Could it be that the use of the term "disease" to describe these changes in plants, which had been known since antiquity, had remained purely metaphorical?

From the beginning of the nineteenth century, a number of practices were put in place that would provide the first shreds of evidence for the existence of pathogenic germs. In 1802, Nicolas Appert (1749–1841)—an *officier de bouche,* or "master of fine dining," for noble families, who later became a self-employed confectioner—set up the world's first food canning factory in Massy, near Paris. The principles of food preservation that he developed involved heating the foods to a high temperature inside a glass container hermetically sealed with a cork stopper. He sold his preserved foods to the French navy. He packaged milk that would keep for two weeks in the same way; today we somewhat incorrectly call it pasteurized milk. It was not until later, when the English replaced glass containers with metal cans, that the invention would take off, thanks to the lower cost and more secure means of transport.

In 1847, Ignaz Semmelweis (1818–1865), a young physician in one of the maternity wards at the general hospital in Vienna, was faced with a high incidence of puerperal fever and deaths of mothers who had just given birth (20%), a rate that was much higher than observed in other maternity wards, and even higher than for at-home births. He linked this abnormally high death rate to the interns' habit of conducting dissections just before taking part in deliveries. By simply having them wash their hands with calcium hypochlorite, a liquid chemically related to bleach, he reduced the mortality rate by more than tenfold. Semmelweis's observation was not well received, however, and was seen as an implicit critique of medical practices. Semmelweis could not provide an explanation for his observations—he believed "cadaverous particles" were responsible for puerperal fever. To correctly interpret the effects of contagion, he would have needed to know that these cadaverous particles reproduced—which he didn't at the time. In his despair, Semmelweis was driven to madness, was committed to an institution, and died from the ill treatment he suffered there.

Like that in cell biology, progress in microbiology was tied to improvements in microscopes and the development of methods for fixing and staining biological specimens.

The work of Louis Pasteur (1822–1895) is emblematic of the transformative effects that germ theory had on medicine, but also on our general understanding of life.[13] While teaching in Strasbourg after having completed his studies at the École normale supérieure, Pasteur's first achievements were the discovery and description of the different isomers (forms) of tartrate. What made this work in chemistry important for that which followed was that it demonstrated the stereospecificity of life—-i.e., that specific isomers were preferentially produced in living organisms. While chemists were incapable of separating isomers, living things did so by synthesizing or making use of only one of a compound's two isomers. When, after being appointed to his position in Lille, his attention shifted toward fermentation and the "diseases" that affect it, he very quickly noticed that the fermentation processes involved a single isomer or led to the formation of a single isomer, which for him was confirmation that living organisms intervened in these phenomena. He revealed the nature of these organisms, and showed that they are specific to a particular type of fermentation. The role of yeasts in alcoholic fermentation had already been proposed by Charles Cagniard-Latour (1777–1859) and Theodor Schwann in

13. Pasteur, 1993.

1838, but the theory was met with opposition from the foremost chemists at the time—Jöns Jacob Berzelius (1779–1848), Friedrich Wöhler (1800–1882), and Justus von Liebig (1803–1873)—who saw it as a return to vitalism.

It was at this point in his career that Pasteur conducted his now-famous experiments on spontaneous generation (1861–1865).[14] The approach to this area of research had barely changed since Spallanzani. Schwann had repeated Spallanzani's experiment in 1837, but reintroduced "purified" air that had passed over heated metal into flasks after boiling. He did not observe any spontaneous generation in the liquid in the bottles under these conditions, though he did demonstrate that the air introduced in this way did not lose its oxygen.

Félix-Archimède Pouchet (1800–1872) published the results of a similar experiment in 1859. He also pumped purified air (purified through mercury) into his bottles, but unlike Schwann he observed that the flasks became cloudy. Faced with these contradictory results, the French Academy of Sciences decided to award a prize to anyone who could carry out experiments enabling the definitive resolution of the question of spontaneous generation. A few months later, Pasteur submitted a series of results in support of germ theory, thus disproving spontaneous generation. He passed air from his laboratory through a cotton filter and showed that it caught particles visible under a microscope that, when added to a flask of clear culture medium, led to the development of multiple microbes. By quickly opening and closing flasks at different altitudes where the air was considered to be more or less pure, he demonstrated that the proportion of flasks that went cloudy varied inversely with the purity of the air. He observed that if nonheated air entered into

14. Farley, 1977.

sterilized flasks through a long tube in the shape of a swan's neck with an inner surface that would catch the germs, the liquid in the flask would not go cloudy and no microbes grew.

The Pasteur-Pouchet controversy has served as a model for many historians of science to demonstrate to what extent the outcomes of scientific debates depended very little on the experiments that were intended to settle them, and rather more on the ideological contexts of the debates and on the social standing of those involved. Indeed, Pasteur never reproduced Pouchet's experiments and thus never demonstrated that the results obtained were incorrect. Pasteur had strong support from the members of the French Academy of Sciences. We will revisit the ideological issues surrounding the controversy in the "Historical Overview" section. However, we can make three observations that are unconnected to this philosophical and religious backdrop. The first is that Pasteur's experimental inventiveness greatly surpassed that of Pouchet, who only reconducted the same experiment over and over.[15] The second is that since Redi, the scope of spontaneous generation had been gradually reduced. At the time of the Pasteur-Pouchet controversy, it involved only the simplest of living organisms. There was increased interest in spontaneous generation for a brief period at the end of the eighteenth century, when preformationism was abandoned and atomist ideas and materialist philosophy gained prominence. However, cell theory, from the beginning but even more so in the form Remak gave it, left no place for spontaneous generation. If the committee of the French Academy of Sciences was satisfied with the experiments carried out by Pasteur, it was because his results were what they had been expecting.

15. Roll-Hansen, 1979.

The final observation is that if this controversy has been recorded in history books, it is because Pasteur came out on top, and his conclusions have since been largely confirmed. If Pasteur had been disproved by later observations, the controversy would probably never have been mentioned again (or even have ceased to exist), or have been brought up only to show how a prestigious institution can get things wrong. This controversy never branched off into other ideas about living things or the origins of life. An analogous debate took place in England in the following years. The assertions of Henry Charlton Bastian (1837–1915) in favor of what was called "heterogenesis," at first favorably received, were soon rejected by the community of biologists, including Darwinians.[16]

Pasteur then studied diseases that affected silkworm farming south of Lyon. In 1835, the Italian entomologist Agostino Bassi (1773–1856) had shown that one silkworm disease was caused by a parasitic fungus, and in 1844 he extrapolated his findings into a general hypothesis to explain the origin of diseases. Pasteur, in parallel with other biologists, identified various diseases, and showed that they were caused by different microscopic germs. He showed how farms could rid themselves of disease by isolating and then breeding uninfected worms and moths. The experience he acquired in this area convinced him that animal diseases could be the result of contamination by specific microbes. Contrary to what the range of his discoveries might suggest, there was a great deal of continuity in Pasteur's work, as revealed by the philosopher François Dagognet. Each stage in his scientific life was a step necessary for accomplishing the following one.[17]

16. Strick, 2000.
17. Dagognet, 1994.

Pasteur is not credited with having isolated the first germs responsible for human diseases. This distinction has gone to the German physician Robert Koch (1843–1910). A simple country doctor and the son of a mining engineer, in 1876 he witnessed the development of an anthrax epidemic in the region where he practiced. This disease primarily affects sheep, but can also be transmitted to humans. The French physician Casimir Davaine (1812–1882) had described the presence of rods in the blood of sick animals and suggested that they were germs responsible for the disease. However, the mechanism of transmission remained a mystery. Koch showed that anthrax bacteria produced spores that could survive many months or even years in the ground before infecting other animals.

Koch's results created a lot of enthusiasm. In 1882, he discovered that the bacterium that today bears his name, Koch's bacillus, was responsible for one of the most dreaded human diseases of the nineteenth century—tuberculosis. One year later he isolated the germ responsible for cholera. In 1880, Karl Eberth (1835–1926) identified the typhoid bacillus, which was isolated by Georg Gaffky (1850–1918) in 1884. The same year, Edwin Klebs (1834–1913) and Friedrich Löffler (1852–1915) isolated the infectious agent responsible for diphtheria. In 1889, Shibasaburo Kitasato (1853–1931) isolated the infectious agent responsible for tetanus.

Koch not only contributed results, but also introduced techniques and a method. He developed an aniline-based dye that allowed microorganisms to be more easily observed. Most importantly, he invented the method for isolating bacteria on a gelatin dish that remains one of the most widely used techniques in microbiology today. He also contributed rules—Koch's postulates—that enabled confirmation that the isolated microbe was indeed the causal agent of the disease. That agent must be present each time the disease is observed, and absent in healthy subjects. The pathogenic germ must be separated from all other microorganisms, and when injected must induce all of the symptoms of the disease in a previously healthy recipient.

Jakob Henle, a student of Johannes Müller, was the first to touch on these criteria theoretically in 1840, and they were also proposed by Klebs.[18] While Henle had understood the need to isolate pathogenic agents from afflicted individuals, he had not considered Koch's final criterion—the possibility of causing the disease by reinjecting the microorganisms isolated from infected persons. Nonetheless, he made two original contributions in his 1840 publication: he stressed the capacity of pathogenic agents to reproduce—which Semmelweis hadn't been able to do—and, more importantly, abandoned notions of germs and ferments, which had been used by Fracastoro, for instance. For Henle, the pathogenic agent was not the seed of the illness—which would imply that they were of the same nature—but the causal agent of it.

During the same period, Pasteur's work would take another direction. In 1880, he discovered by chance that an old culture of microbes responsible for chicken cholera had lost its pathogenicity, but protected the birds from being infected by virulent forms of the same germs. He deduced that the culture environment had somehow reduced the pathogenicity of the microbes. He then used this attenuation principle to develop a vaccine against anthrax, whose effectiveness in sheep was confirmed through a spectacular experiment carried out in 1881 in Pouilly-le-Fort. In 1885, he applied the same principle to cure a child bitten by a rabid dog. This method would gradually be expanded

18. Brock, 1999.

to include all infectious diseases whose pathogenic agents or the toxins they secreted could be at least partially isolated. Some microbes are pathogens owing to the toxins they produce, as demonstrated by Émile Roux (1853–1933) and Alexandre Yersin (1863–1943) in the case of diphtheria.

The development of serotherapy would be as rapid as that of vaccination a few years earlier. In 1890, Emil von Behring (1854–1917) and Kitasato demonstrated that an animal injected with a nonlethal dose of the diphtheria toxin would produce antitoxins in its blood (now known as antibodies), which, when injected into another animal, could prevent or cure the illness. The experiment was reproduced for tetanus and anthrax, and Roux performed it on human subjects from 1894 in Paris at the Necker hospital.

The practice of variolation, which came from India, China, and Africa, and would be employed in Europe at the beginning of the eighteenth century on the initiative of Lady Mary Montagu (1689–1762), in many ways anticipated Pasteur's findings. It consisted of inoculating individuals against smallpox by injecting them with liquid from the pustules of recovering patients. Despite some deadly accidents, the effectiveness of variolation was demonstrated by the mathematician Daniel Bernoulli in 1760: those who had undergone it saw their life expectancy increased by three years. Vaccination was developed by Edward Jenner (1749–1823) in 1796. The practice resulted from his observation that individuals who tended cows would sometimes contract a benign disease from the animals that resembled smallpox, but that they never contracted smallpox itself. Jenner systematically inoculated children with pustules from cows suffering from cowpox, and demonstrated that it protected them from smallpox as a result.

What was new in the practices of followers of Pasteur, however, was the widespread use of the process and its explanation.

For Pasteur, the attenuated pathogenic agent competed with the infectious germ to obtain a substance X found in the organism (whose nature was unknown) that was needed for its growth. Thus, it prevented the development of the virulent microbe. An immunological explanation, via the synthesis of antibodies that recognize both the virulent germ and the weakened germ, would be proposed only a few years later.

Pasteur's accomplishments were covered extensively in the media, primarily because of his strategy. Rabies was an excellent choice—not because the disease was a public health issue, but because the way it killed people was terrifying. After the first successes were confirmed, people bitten by rabid animals poured in from around the world, from Russia to the United States, to be vaccinated.

All of these discoveries resulted from fierce competition between microbiologists, which was compounded by the German-French rivalry. And so, the same year Koch identified the cholera bacterium, Pasteur had sent a team to Egypt with the same objective. Not only was this expedition a scientific failure, but young Louis Thuillier (1856–1883) lost his life there. This competition accounts for the rapid accumulation of discoveries. In addition to those already described, there was also in the same period the identification by Yersin of the bacillus responsible for the plague, during an epidemic that had broken out in Hong Kong, as well as that of the agent responsible for syphilis, the spirochete *Treponema pallidum*, by Fritz Schaudinn (1871–1906) in 1905.

The development of vaccines occurred in a similar way, though at a slower pace. In 1892, Waldemar Haffkine (1860–1930) of the Pasteur school prepared a vaccine against cholera, and in 1897 a vaccine against the plague. In 1896, Almroth Wright (1861–1947) developed the first effective vaccine against typhoid fever.

Germ theory also provided the means to protect oneself against infectious diseases through the use of antiseptics. Joseph Lister (1827–1912), who drew inspiration from Pasteur's ideas, was the first to implement the concept, using phenol. Koch developed other methods for antisepsis, and the effect of these protective measures could henceforth be directly ascertained by observing how they acted on the relevant microbes.

However, the Koch and Pasteur schools did not share the same view of the causal role of microbes in diseases. In Koch's warlike view, ridding oneself of a disease was achieved through the eradication of the microbes. This contrasted with a different view of disease as proposed by one of Pasteur's students, Émile Duclaux (1840–1904),[19] and later by Albert Calmette (1863–1933) and Charles Nicolle (1866–1936), whereby the fight against a disease involved strengthening an organism's defenses as much as it did eradicating the pathogens involved. However, the two schools shared the same conviction that each disease was linked to a specific microbe whose characteristics were stable. Ferdinand Cohn (1828–1898) would undertake the classification of these microbes. Both schools were against the theories on the pleomorphic nature of microbes supported by some biologists, such as the botanist Carl von Nägeli (1817–1891), whereby more than one distinct form could exist.

The microbial revolution led to the creation of a new type of scientific institution, one that combined facilities for research on infectious diseases, a teaching center for the new discipline, and facilities for the production of vaccines and serums. The Pasteur Institute was established in Paris in 1887, the Institute for Infectious Diseases in Berlin in 1891, the Lister Institute in London in 1893, and the Rockefeller Institute in New York in

19. Morange, 2006b.

1902. Each of these institutions took a slightly different approach. The Berlin institute specialized in medical bacteriology, the Pasteur Institute focused its efforts on describing pathogenic and nonpathogenic microorganisms, and the Rockefeller Institute became a center for experimental medicine. The creation of these institutes represented a separation from universities and a shift in biological research toward medical objectives.

The discovery of pathogenic germs was one of the first clear signs heralding the globalization of modern science. Pasteur founded Pasteur Institutes overseas in French colonies (and in a few other countries), with the goal of enabling those countries to benefit from progress in microbiology. This initiative led to the identification of a new category of disease—tropical disease—which was in most cases caused by protozoans, not bacteria. The agent responsible for malaria was characterized by Alphonse Laveran (1845–1922) in 1880 in Algeria, and in 1897 Ronald Ross (1857–1932) described the complete life cycle of the pathogen in the mosquito (anopheles) and its human host. There followed descriptions of sleeping sickness and Chagas disease, leishmaniasis, and many of the other illnesses that had made Africa in particular the "white man's tomb" in the minds of most French people.

Physiology's Golden Age

The nineteenth century is often considered to be the golden age of physiology. This is despite the experiments conducted by Erasistratus in Alexandria that belonged to the realm of physiology, and the work of Harvey and Borelli in the seventeenth century and Haller and Spallanzani in the eighteenth century, all of whom were physiologists. One should probably see in this characterization an acknowledgement of the important role that physiology played as a model in the reorganization of German universities and of universities worldwide, wherein teaching and research became closely intertwined.

However, in the nineteenth century physiology was far from being a unified discipline. We can distinguish four major areas of research, each with its own methods and questions, although they occasionally intersected.

The first area, which best represented physiology in the eyes of outside observers, is represented by the work of François Magendie (1782–1855) and Claude Bernard (1813–1878).

Magendie, a professor at the Collège de France, was a supporter of the experimental method (and of vivisection, which is one specific form of it). He was partial to facts and suspicious of theories. In contrast to Bichat, he was convinced that living phenomena were just physical and chemical phenomena. He is known for his work on circulation and respiration, but primarily, following the work of the English physician Charles Bell (1774–1842), for having identified the sensory and motor roots of spinal nerves. He and his work were, however, eclipsed by the man who was his laboratory assistant at the Collège de France, and who succeeded him as chair: Claude Bernard.

Bernard owed a great deal to Magendie—he accorded the same importance to the physicochemical approach to life and to the experimental method.[20] Much of Bernard's work includes emblematic examples of this method. Using the experimental method, he showed that the liver played an essential role in regulating blood sugar levels by storing sugar from food as starch (which today we would call glycogen), and could even make sugar from other sources beyond food, and redistribute

20. For the work of Bernard, see Grmek (1973), Holmes (1974), and Duchesneau et al. (2013).

it to the organism between meals. Bernard thus reaffirmed the central role of the liver in nourishment, which had been asserted since antiquity but seemed to have been called into question by the discovery of blood circulation and of the lymphatic system. One of his first projects was to reproduce Van Helmont's results on the distinct roles of gastric acid and ferments in the stomach's digestive process. He also described the role of the pancreas in digestion and shed light on the action of certain poisons, such as curare.

Bernard did even more—he described his method in *Introduction à l'étude de la médecine expérimentale* (*Introduction to the Study of Experimental Medicine*),[21] and proposed a new view of life in his posthumous work *Leçons sur les phénomènes de la vie communs aux animaux et aux végétaux* (*Lectures on the Phenomena of Life Common to Animals and Plants*) (1878–1879),[22] as well as new ideas on the connections between physiology and pathology.

It is difficult to assess the impact of his method. It was received more enthusiastically by philosophers (and nonscientists) than by his colleagues in biology. The method was derived entirely from his own research activity. Of the 16 experiments he used as models, 15 were his own—which does not, however, mean that he described them faithfully. Indeed, in his notebooks the descriptions and order of steps for these experiments are quite different.[23] His book is an exposé of the hypothetico-deductive method—the triad of observation, hypothesis, and experiment—though with some original features. Bernard believed that some experiments could serve solely as means for

21. Bernard, 2018.
22. Bernard, 1974.
23. Grmek, 1973.

exploring the unknown and uncovering facts. Hypotheses played an important role in his process, and in his view developing them involved mental activity similar to that of artistic creation. However, unlike Pasteur, he thought hypotheses had only limited truth value. In the advancement of scientific knowledge, he compared hypotheses to horses that are successively exhausted, and, paradoxically, he believed that there was no scientific method that could be applied in all places and in all circumstances.

Claude Bernard's conception of life was also complex and often misunderstood. With his insistence on the determinism of physicochemical phenomena in living things and his rejection of vitalism, he is often considered a materialist proponent of a reductionist approach to biological phenomena. Indeed, he equated vitalism with materialism—two metaphysical theories that in his view had no place in science. He was convinced of the uniqueness of living phenomena, which he saw as operating on two levels—in the capacity of organisms to perform chemical syntheses, and in their capacity to develop. Bernard thought that the chemistry of the living, that which builds the components of life, could not be compared to that practiced by organic chemists, and was connected to the matter specific to living beings—protoplasm. He admitted that cells existed and played a role, but for him protoplasm was paramount—the formation of cells only gave it its form.

Finally, Bernard provided a new, quantitative view of pathological phenomena. He distinguished three types of living things: those that are totally dependent on environmental conditions, suspending their vital activity when conditions are unfavorable; those that are simply affected by environmental conditions and modify their activity as a function of environmental variations; and those that possess a constant internal environment that is unaffected by external variations. For the latter, any significant deviation from normalcy in the internal environment constituted a pathological phenomenon. The pathological was therefore nothing more than a quantitative change in physiological parameters, an opinion already expressed by François Broussais (1772–1838).

The second area of work in physiology is more "chemical," in the sense that the researchers' work was focused directly on the chemical characterization of the components that make up life.[24] It is often difficult to distinguish this work from that of chemists. This line of research was first pursued by Lavoisier and Priestly; there were also the Swedish chemist Carl Scheele (1742–1780), who characterized lactic acid, citric acid, and uric acid, and the French chemist Antoine-François Fourcroy (1755–1809), Lavoisier's friend and successor, who systematically described the chemical components of life. Similarly, in 1806–1808, the Swedish chemist Jöns Jacob Berzelius published a treatise, *Lectures on Animal Chemistry,* in which he described the chemical composition of blood, milk, bones, fat, and so on. He coined the term "protein" to designate a collection of nitrogenous substances whose presence in organisms was described at the beginning of the eighteenth century. His essential contributions were the introduction in chemistry of a new form of notation and of the notion of catalyst—a substance that speeds up a reaction without participating directly in it. He identified ferments as catalysts. An early ferment, amylase, which catalyzes the conversion of starch into maltose, was isolated in 1833 by the French chemists Anselme Payen (1795–1871) and Jean-François Persoz (1805–1868), who were working for the sugar industry. In 1877,

24. Fruton, 1999; Tanford and Reynolds, 2001.

the German physiologist Wilhelm Kühne (1837–1900) would name these biological catalysts "enzymes."

Berzelius's student Friedrich Wöhler is known for having carried out the chemical synthesis of urea, an organic molecule, and thus demonstrating that there was no difference between the chemistry of living beings and ordinary chemistry. However, this interpretation of Wöhler's experiment came only much later. Nonetheless, it is true that all of these chemists shared the same suspicion of hypotheses, and had a very limited interest in cell theory; instead, they adopted the distinctions set out by Bichat between the different tissues in an organism and the chemical processes that occur within them, while rejecting Bichat's belief in vitalism. They shared the same conviction that precise descriptions of the molecules of living things would pave the way to understanding their functions.

We have already mentioned Liebig, the first chemistry professor to have a research laboratory in a German university. Liebig was convinced that fermentation was a purely chemical phenomenon, and, for that reason, he disputed Pasteur's results. He played a major role in the promotion of chemical knowledge for the purpose of increasing agricultural production—the beginnings of agrochemistry.

Félix Hoppe-Seyler (1825–1895), a professor in Tübingen and then in Strasbourg, was the best example of what he, himself, would have called a physiological chemist. He demonstrated that oxygen is transported in the blood by hemoglobin, a protein present in red blood cells, which he was able to crystallize and to demonstrate that it contained iron. He also characterized chlorophyll. In 1877, he founded the journal *Zeitschrift für physiologische Chemie* (Journal of physiological chemistry) and attempted, without success, to establish a chair in physiological chemistry in order to have this new approach to the living world recognized as a discipline.

In 1869, one of his students at the University of Tübingen, Friedrich Miescher (1844–1895), isolated a substance he described as "nuclein" owing to its being found in the nucleus of cells. As biological material for this extraction, Miescher chose pus, particularly rich in leukocytes, white blood cells, whose nucleus makes up nearly all of the cell's volume. He gathered the cells by washing bandages from sick patients. The results were so surprising—the chemical composition of this substance differed from all other known biological substances—that Hoppe-Seyler did not allow him to publish his results until 1871, after having reproduced all of the experiments himself.

Another student of Hoppe-Seyler, Albrecht Kossel (1853–1927), continued the work carried out by Miescher. He showed that nuclein was made up of proteins and nucleic acid. He characterized the five bases—adenine, thymine, guanine, cytosine, and uracil—present in this acid. Kossel also studied proteins, characterized several of the amino acids that make them up, and proposed the hypothesis that proteins are formed from chains of these amino acids.

Emil Fischer (1852–1919) would add to Kossel's work on proteins by introducing a new method enabling an accurate measurement of the proportions of each amino acid in a protein, and by determining the nature of the peptide bonds between amino acids; this would allow him to synthesize the first artificial peptides (containing up to 18 amino acids). Fisher is also known for having described the interaction of enzymes with the molecules they transform as being akin to that of a lock and key (1899–1908). He also contributed to a better understanding of the structure of nucleobases and sugars.

Plant physiology developed a little more slowly, drawing from the two preceding traditions of animal physiology. The German physiologist Julius von Sachs (1832–1897), working in Würzburg, was certainly the researcher who best embodied this new approach to the study of plants. He very quickly subscribed to Darwin's theory and pitted himself against both the classifiers and the followers of the morphological tradition that drew inspiration from *Naturphilosophie*. Plant physiology had to involve experiments and required the precise, quantitative study of phenomena. It also involved chemistry: he deciphered the syntheses that accompanied photosynthesis and, like Liebig, characterized precisely the nutritional needs of plants.

Just as chemists were playing an increasingly important role in the study of living things, so were physicists, who were regaining a foothold after the setbacks of iatrophysics. The discovery of "animal electricity" provided them with an opportunity to reenter the field of biology. The Italian anatomist Luigi Galvani (1737–1798), working in Bologna, was behind this discovery. However, the stage had been set by observations made in both South America and the Netherlands on electric fish, such as torpedo rays.[25] In 1781, Galvani observed that the muscles of a dead frog would contract when jolted by an electric spark. From this he deduced that muscle movement was caused by an electric fluid transported by nerves, which he called "animal electricity." Contrary to what had been thought since antiquity and up until Swammerdam, nerves did not transport a physical fluid but an electric one.

The physicist Alessandro Volta (1745–1827) reproduced Galvani's experiment and confirmed his results, though his interpretation of them was quite different. He did not believe that the

25. Koehler et al., 2009.

electricity was of animal origin, but rather that it was produced as a result of different metals being put in contact with the frog's body. To confirm his interpretation, Volta constructed the first electric battery in 1800, and based on his description of what we would now call the redox potential of different metals, he provided the general blueprint for building one. The manufacture of batteries, which represented a considerable technological advancement, has its roots in a scientific controversy in which both protagonists were right (and wrong): animal electricity did not cause the contraction of the frog's muscles as described by Galvani, but animal electricity did indeed exist.

Going beyond and correcting Lavoisier's work, in the 1840s, Julius Robert Mayer (1814–1878), the physicist James Joule (1818–1889), and the biophysicist Hermann von Helmholtz (1821–1894),[26] a student of Johannes Müller, established the relationship between work and heat, thereby laying the foundations for the first principle of thermodynamics on the conservation of energy.

In Berlin, Helmholtz made a number of discoveries in the fields of physics and mathematical physics. In physiology, he was the first to measure the speed at which nerve impulses propagated. He worked on the physical mechanisms for perceiving sounds and images, and researched the physical bases of musical aesthetics. While his hypothesis that different colors were perceived separately turned out to be true, his theory of hearing based on resonators—that individual cells in the cochlea resonated at a specific frequency of sound—would have to be corrected.

Another important player in physiological physics was Emil du Bois-Reymond (1818–1896), who was also a student of

26. Meulders, 2010.

Johannes Müller. He described the nerve action potential, whose speed of propagation had been measured by Helmholtz. He was convinced that animal electricity was a result of "electric molecules" in tissues such as muscles. However, despite his reductive tendencies, he was somewhat equivocal about the value of scientific knowledge. He thought that certain questions had not found and would never find answers in science, including such things as the nature of matter and forces, and the origins of movement and simple sensations.

The fourth area of physiology dealt with locating higher mental functions in the brain, and, more specifically, the cerebral cortex.[27] It came out of the work of anatomists and physiologists as a major question in science that joined, more or less successfully, the efforts of physiognomists to discern individuals' qualities from the shapes of their faces. Physiognomy is an ancient "science" (some of its first writings have been attributed to Aristotle) that had some success throughout the Middle Ages and the Renaissance. Painters such as Charles Lebrun showed a great deal of interest in it.

Since Galen, there had been many attempts to localize higher cognitive functions, and speculation would gradually be replaced with precise experiments. Pierre Flourens (1794–1867) demonstrated the role that the cerebellum played in movement. However, the discovery that made the most lasting impact was that of the "seat of speech" by Paul Broca (1824–1880) in 1861. While treating a patient with aphasia, and after examining the patient's brain following his death, Broca connected the loss of speech to a lesion in a specific area of the left frontal lobe, which has since become known as Broca's area.

27. Clarke and Jacyna, 1987; Finger, 1994, chapter 3.

Apart from a few other anatomic studies of the limbic system and the rhinencephalon, or olfactory brain, the bulk of Broca's work was anthropological. In 1859, he founded the Société d'anthropologie de Paris (Paris Society of Anthropology); in 1872, the *Revue d'anthropologie*; and in 1876, the École d'anthropologie de Paris. Broca was convinced that the size of the skull (and the brain) were directly connected to intelligence, a conclusion he supported by the different weight measurements he obtained for the brains of men and women, and those of Europeans and Africans.

Broca made an unsuccessful attempt at creating the first functional brain imaging apparatus, which he called a "thermometric crown." The idea behind it was to measure the temperature of different areas of the brain through the skull: the more active an area of the brain, the warmer it was. Broca was also the founder, in 1848, of the Société des libres penseurs (Free Thinkers' Society).

Broca's observations that led him to the hypothesis of a seat of language in the brain provoked a heated debate between those like him who thought functions were localized and holists, like Flourens, who were convinced that the cerebral cortex functioned in an integrated way. The first camp found support in the demonstration by David Ferrier (1843–1928) of the existence of a motor cortex in monkeys, and confirmation in the electrophysiological observations of Richard Caton (1842–1926) in Liverpool. However, the shadow of Franz Joseph Gall (1758–1828), a doctor who worked in Vienna and later in Paris, hung over the debate. An excellent anatomist, Gall was convinced that intelligence and character traits were located in the cerebral cortex, and he believed that he could deduce a person's virtues and faults by examining the exterior of the skull, whose shape reflected that of the brain housed within it. Like most biologists, Flourens had voiced his negative opinion of this new science of phrenology.[28] However, phrenology had generated considerable interest among the general public, continuing the earlier interest in physiognomy. The Italian physician Cesare Lombroso (1835–1909) built on Gall's observations when he stated in his 1876 work *L'uomo delinquent* (The criminal man) that it was possible to identify natural-born criminals from their physical appearance, and particularly from the shape of their skulls.

The debate between localizationism and holism would be taken up again at the end of the century, though in a different form, between the Italian biologist Camillo Golgi (1843–1926) and the Spanish biologist Santiago Ramón y Cajal (1852–1934). Golgi asserted that the brain was a continuous network, while Ramón y Cajal thought that nerve cells, or neurons, were separated by a microscopic space at the point of contact, which the physiologist Charles Sherrington would call the "synapse." This time it was not a debate between two researchers using different approaches and techniques, which is often the case in science and makes the results more difficult to compare. Ramón y Cajal used the silver staining of neurons, a technique developed by his "adversary." The Nobel Committee made the prudent decision not to choose between neuron theory and reticular theory, and in 1906 awarded the Nobel Prize to both Golgi and Ramón y Cajal. Neurophysiologists in the twentieth century settled the debate in favor of Ramón y Cajal and neuron theory. That victory was effectively a victory for cell theory in a field where dividing an organism's functions among multiple cells seemed the most problematic.

28. Lantéri-Laura, 1970.

Historical Overview

The Roots of Cell Theory

As we have seen, trying to pinpoint the genesis of cell theory is an almost impossible task for historians. Not only have cells been described in different ways on multiple occasions by a number of writers since the middle of the seventeenth century, but the date that has made its way into the history books refers to a form of cell theory that is quite different from the one that is taught today.

The difficulty of the emergence of cell theory can also be attributed to the extended period of time over which it occurred. With the accumulation in the early nineteenth century of observations demonstrating the presence of cells in various tissues, the idea became more mainstream, and a contrario the more difficult it became to propose a cell theory. The discovery of microbes, and of an evolution of living organisms, also endured a slow normalization before the ideas were accepted.

It is therefore tempting to see the rise of cell theory as resulting from events that were unrelated to the pursuit of scientific knowledge. For example, Johannes Müller's influence and his support of Schwann and Schleiden's theory are thought to have been key factors in its broader acceptance. More generally, the great influence of German science meant that biologists in Germany may have been unaware of the work being conducted elsewhere. The way in which Schwann presented his theory would also have played an important role: he explicitly asserted, much more vigorously than most biologists, that cells were "the elementary particles" of all animal tissues.

Another explanation for the genesis of cell theory can be found in the philosophical ideas of the period, in its "episteme."

Georges Canguilhem saw two starting points for cell theory.[29] The first was in Buffon's work and in his ideas on organic molecules. Canguilhem had only to make the conceptual leap from organic molecules to cells to make cells "Buffon's dream." The second for him was German *Naturphilosophie*, and, more specifically, the work of the biologist Lorenz Oken (1779–1851). Adherents of *Naturphilosophie* sought to draw up a general theoretical structure for nature in order to establish the foundations of the natural sciences; cell theory was perfectly suited for this. Oken was not just a naturalist; he was also one of the masters of *Naturphilosophie*. In 1805, in a work on procreation, he asserted that all living things were born from cells and were made up of cells. He also discussed the individuality of cells and their relationship with the organism of which they are part. The discussion evokes an analogy with the organization of human society and the relationship between the individual and the state.

Separate from the more general discussion on the role of philosophy in the development of scientific knowledge, however, these types of explanation provide little to account for the 90 years that separate Buffon's model and Schwann's theory, or the 40 years that were apparently needed for Oken's ideas to gain influence.

Henry Harris has put forward a hypothesis that is more in line with this chronology.[30] In his view, the key event was the discovery of the cellular nucleus and the nucleolus, which, as we have seen, Schwann believed to be at the origin of cell formation. Thanks to these discoveries, which occurred in the 1830s, cells gained an organizational center and a structure, and

29. Canguilhem, 1965c.

30. Harris, 1999.

were no longer simple sacs found within organisms. This hypothesis is interesting for three reasons. First, it retains the chronology of discoveries and explains why cell theory developed quickly following the discovery of these organelles. Secondly, it also explains why the nucleus has continued to play a central role. Thirdly, one cannot help but compare the role of the nucleus in cells to that of the sun in the solar system, which controls the movements of all the planets through its gravitational force. This is another example of the influence that physics, and in particular Newtonian physics, has had on biology since the middle of the eighteenth century. It was this combination of factors, and in particular the enthusiasm around the discovery of the nucleus and nucleolus, that enabled the acceptance of a cell theory that had already become mainstream in a weakened form.

As we have seen, the difficulty in constructing a cell theory was also due to the fact that, for many biologists, and in particular physiologists, it contributed nothing, and only complicated an explanation of the chemical phenomena occurring in living things.

Scholars Trapped by Their Own Philosophical Ideas?

The first half of the nineteenth century is often presented as a prime example of a period where philosophical ideas about the world were able to stimulate and direct the development of the sciences. *Naturphilosophie,* of which Johann Gottlieb Fichte (1762–1814), Friedrich Wilhelm Joseph von Schelling (1775–1854), and Georg Wilhelm Friedrich Hegel (1770–1831) are considered to be the fathers, would have provided a framework within which the natural sciences could develop. It could be argued that the development of cell theory was supported by the desire to find a unifying principle of organization within the living world, in agreement with the objectives of *Naturphilosophie*. Goethe (1749–1832), who was a writer and politician but also a promoter of *Naturphilosophie* and a naturalist, made important scientific contributions. In his 1790 work *Metamorphosis of Plants*, he was the first to demonstrate that the leaf represented the basic archetypal form for plant structures, and that flowers are modified leaves. More generally, the success of German physiology is seen as the outcome of that philosophical tradition.

However, there are several objections to raise to this hypothesis. The first is that *Naturphilosophie* was not the only philosophical worldview that German scholars adhered to. Some historians believe that Kant's philosophy and that of his followers, such as Blumenbach,[31] played a more important role. This influence can be seen in the efforts of a number of biologists to precisely describe the organization of living things and to account for teleological phenomena in the living world without resorting to religious explanations. The search for "vital materialism," for a chemistry that was unique to the living, was another facet of the influence of Kantian philosophy. It is to this philosophical tradition, much more so than to *Naturphilosophie*, that we should attribute the birth at the end of the eighteenth century of a science specifically of the living world—biology—a term that in 1802 appeared both in the work of Gottfried Reinhold Treviranus (1776–1837) and in that of Jean-Baptiste Lamarck (1744–1829).

The second difficulty is in making these philosophical models "stick" to particular theories or scholars. Though French scholars are credited with playing a role in the development of

31. Kant, 2007; Lenoir, 1982.

cell theory, that was materialists like Buffon and Raspail, and not the adherents of a philosophy that was more or less spiritualist.

We run into similar difficulties in pinpointing the philosophical position of the German scholars who participated in the genesis of cell theory. We have seen the major, but indirect, role that Johannes Müller played through the many scholars he rubbed shoulders with, and the fact that he was quick to support this new theory. He is often thought to have been one of those scholars whose work was influenced by their philosophical training. He later rejected *Naturphilosophie*, but never abandoned a certain form of vitalism. However, despite his influence, his students adopted wide-ranging philosophical positions, from spiritualism to militant materialism.[32]

It has been suggested that the paleontologist Georges Cuvier inherited his ideas about organisms from *Naturphilosophie*, which he would have encountered during his early years at university (see next chapter). However, Cuvier adamantly rejected Goethe's and Oken's "discovery" that the bones of the skull developed from vertebrae. In a well-known debate between Cuvier and Étienne Geoffroy Saint-Hilaire (1772–1844), it was Saint-Hilaire, a supporter of materialism, who was closest to the ideas in *Naturphilosophie*.

Similarly, Bichat is often portrayed as a spokesperson for ideological philosophers such as Georges Cabanis (1757–1808), a close friend of his, and Étienne Bonnot de Condillac (1715–1780). How do we then reconcile these two philosophers' materialist philosophy with Bichat's vitalism? We must also be wary of claims of supposed influence. That Schleiden studied at the University of Jena "in the shadow of Oken," as Georges Can-

32. Otis, 2007.

guilhem put it,[33] and that Schwann had been a student of Johannes Müller, who had been influenced by *Naturphilosophie* in his youth, might appear persuasive (even though the expression "in the shadow of" is not exactly precise), but this influence is mere assumption; that cell theory was a product of *Naturphilosophie* remains to be demonstrated.

Philosophical concepts are often used by nonphilosophers—scholars and scientists—as rallying calls, and more often as shields, in their debates, which for an outside observer makes the relationship between philosophical ideas and scientific models all the more opaque. Thus, Liebig was thought to have referred to *Naturphilosophie* as "a plague." Claude Bernard positioned himself as a champion in the fight against vitalism, even though his own ideas on the uniqueness of living phenomena were much more ambiguous. Scientists' repeated references to Leibniz and Spinoza must therefore be taken with a grain of salt, given their feeble knowledge of philosophy. The most enlightening example of this is surely Pasteur. His rejection of any form of spontaneous generation, and the great deal of proof he supplied in support of the uniqueness of the chemistry of living things, stemmed from his religious convictions. In a lecture given at the Sorbonne in 1864, he asserted that if spontaneous generation did indeed exist then so did the evolutionary changes in organisms, and the notion of a Creator became pointless. However, we should not forget that Pasteur conducted multiple (unsuccessful) experiments seeking to demonstrate that the unique chemistry of living beings could be reproduced in vitro with the right environmental conditions, such as through the use of magnetic fields. Moreover, the extent of Pasteur's religious convictions is not known. André Lwoff has recounted what Pasteur revealed in

33. Canguilhem, 1965c.

confidence to Olga Metchnikoff just prior to his death—that he was not a believer, but chose not to break with religion out of respect for convention.[34]

But these reflections do not conflict with the idea that an important shift in attitude occurred at the end of the eighteenth century, with the development of romanticism and the discovery of "sensitivity"—a concept that was closely associated with *Naturphilosophie.*[35] German biologists' lack of interest in vivisection is one sign of this. And it is not a coincidence that homeopathic medicine was established in 1807, through the initiative of the German physician Samuel Hahnemann (1755–1843). Mechanistic explanations had lost their supremacy. The discovery of electrical and magnetic phenomena and their assumed importance in the functioning of living things contributed to this. Descriptions of cells and cell division were not mechanistic, and would become so again only in the second half of the twentieth century, following the rise of molecular biology.

But the importance of vitalism also diminished at the beginning of the nineteenth century, becoming only what François Jacob said about holism in *The Logic of Life*—a defense against hasty and simplistic explanations of living phenomena.[36]

Scientific ideas are connected to but are not dependent on the culture of their times. Building scientific knowledge depends above all on the techniques used and traditions of research that can persist over long periods of time. The "attitude" of scientists is important, but similar attitudes can stem from very different philosophical and religious beliefs. In his quest

34. Lwoff, 1981, pp. 43–64.
35. Richards, 2002; Gaukroger, 2010.
36. Jacob, 1993.

for well-established facts, a nineteenth-century organic chemist is probably closer to an empiricist believer such as Cuvier than to a visionary materialist such as Geoffroy Saint-Hilaire. Similarly, the desire to find a single explanation or model that explains all phenomena is a mind-set more often resulting from the training one has received than from an adherence to some philosophical theory.

The Tension between Chemical Explanations and Structural Models

As in previous centuries, chemical explanations of living phenomena in the nineteenth century existed alongside structural ones. The ease or difficulty of bringing together these types of explanations depended on the structural system proposed. Bichat's tissue model goes well with chemical explanations of life, because each tissue type is supposed to have a unique chemistry. This probably partly explains the longevity of Bichat's model and the fact that Claude Bernard devoted several pages to it in his last book, while hardly mentioning cell theory.

Indeed, cell theory is problematic in that it fragments the chemistry of living things in seemingly unnecessary ways. The success of the concepts of cytoblastema, sarcode, and protoplasm, all of which refer to matter that carries the chemistry of life, bears witness to the need to get around the problem—although at the risk, according to Schwann and Claude Bernard, of turning the cell into only a secondary product of that living matter.[37] It was only in the twentieth century that chemical and structural descriptions of life would come together.

37. Loison, 2012.

Was Embryology Holding Out for Evolution?

Baer's embryologic model and, even more so, the model proposed by Meckel and Serres seemed to anticipate the arrival of Darwin's theory of evolution, given that the theory justified them so well. However, this is an illusion that stems from a retrospective viewpoint. Both models were conceived relative to the great chain of being, whereby organisms are classified from the simplest to the most complex.

The mismatch between these models and an evolutionary concept of life can help us understand that there are two very different ways of thinking about evolution. To use modern terminology, evolution can be seen as a program running its course, an inexorable march toward complexity. This is Lamarck's view, which was also shared by Spencer and Haeckel. Evolution can also be thought of as a succession of chance occurrences resulting from historical and thus contingent events, the emergence of new things. This is the door that Darwin would open. In the first case, evolution is simply the unfolding of the complex chain of being over the axis of time, and ideas based on the existence of a chain of being can easily be incorporated into it.

1859: A Remarkable Year

For the field of biology, 1859 was the equivalent of 1905 for physics: a year when three events that were independent and seemingly contradictory, but in fact complementary, would occur. Darwin published *On the Origin of Species*; Pouchet published a book on heterogenesis (spontaneous generation), which led Pasteur to undertake his famous experiments on spontaneous generation; and Virchow would define cell theory in its final form. Darwin forced one to think about the origins of life in nonliving matter, while Pasteur seemed to introduce a radical distinction between the two. However, this opposition is only apparent: to be able to conceive of a *true* history of life, with a beginning, this history could not repeat itself ad infinitum. In a counterintuitive way, Pasteur's work was as important to the advent of a historical view of life as Darwin's. As for cell theory, it put a "logical" end to the possibility of spontaneous generation. When the mechanism for fertilization would be described a few years later, it would make the history of life a history of cells and cell division.

Contemporary Relevance

The many questions that biologists were asking in the nineteenth century seem familiar to twenty-first-century biologists, though the terminology may have changed. The relevance of a systemic view in explanations of living things is reminiscent of the debates in the nineteenth century on the localization of functions within organisms. Similarly, the role of mathematics in describing biological phenomena occurs in twenty-first-century discussions much as it did in the nineteenth century. Following Auguste Comte, Claude Bernard granted mathematics a limited role, and was particularly opposed to the use of statistics. Also in the nineteenth century, and owing to the increasing specialization of biological disciplines, there gradually developed a divide between a biology based on descriptions of structures and functions and one that was more interested in the description of organisms and their relationships with the environment. This occurred despite the unifying effect of both cell theory and the theory of evolution. The division persisted through the first half of the twentieth century, and has only recently begun the difficult process of being resolved.

The Disappearance of Traditional Disciplines in Biology

As we have seen, physiological work became more and more varied. Hoppe-Seyler sought to acknowledge this by creating a new discipline that he called "physiological chemistry," and though he was unsuccessful, the establishment of this discipline would be achieved by biochemists at the beginning of the twentieth century.

Molecular biology finds itself in the same situation today as that faced by physiology at the end of the nineteenth century: it is threatened in all directions by the creation of new disciplines—systems biology, synthetic biology, epigenetics, and so on. Will it disappear, or at least lose its dominant position, as was the case with physiology in the twentieth century?

But what should we make of such a disappearance? A lot of research conducted in other disciplines today is an extension of the experimental tradition of nineteenth-century physiology. The creation and disappearance of disciplines seems hardly connected to the continuity of research programs, or to the similarity of experimental approaches. Which makes the category of "disciplines" a rather irrelevant tool when it comes to the history of ideas.

The Endogenous or Exogenous Origins of Diseases

Virchow's cellular pathology contrasted with the germ theory of Pasteur and Koch. But there was dialogue between these two approaches to pathological phenomena: the development of a metastatic growth in previously healthy tissue was compared to an infection from external germs. However, in cell pathology, diseases originate at least in part inside the organism, while in germ theory their origins are completely external. This probably explains why Virchow objected to the observations of Semmelweis, when questioned about the precise role of microbes in disease, and showed little interest in the development of antitoxins. This separation did not exist in Hippocratic medicine, but once it was clearly introduced in the nineteenth century, it would not disappear. Remarkably, the same opposition occurred much earlier, concerning plant diseases.[38] It also underlies the succession of twentieth-century models that have attempted to explain cancer. Even more recently, the causal role of the HIV virus in AIDS has been contested by some biologists, such as Peter Duesberg.

As we have seen, a related opposition separated the schools of Koch and Pasteur. Koch aimed to eradicate diseases by eliminating the germs responsible for them. Pasteur, and even more so his disciples, sought to weaken pathogenic agents and to reinforce an organism's defenses. With many infectious diseases, the pathogenic germ is always present in the environment and the disease breaks out only when conditions are favorable for it. This is the same observation made by botanists in the eighteenth century when they saw that a benign plant parasite could become a pathogen in certain circumstances.

This debate is still alive and well today. After having wiped out smallpox, should other diseases such as poliomyelitis be eradicated? Such a policy to eliminate a virus through vaccination runs into two difficulties. The first is actually achieving the goal, as the disappearance of a virus from the last affected zones

38. Denis, 2011.

is always difficult. The second is knowing what to do with the pathogenic agent and its weakened forms after the disease is eradicated. Should we also get rid of them, or preserve them to be able to deal with the virus's possible return? These questions often play out in books and films. As with other seemingly simple ideas, eradication is often presented as the only good solution.

The Debate on Cerebral Localization

This debate is as lively today as it was in Flourens's and Broca's day, though it has shifted somewhat. In the nineteenth century, the issue at stake was knowing whether one could deduce, from the connection between an alteration to an area of the brain and a loss of higher function, that a particular cerebral area was responsible for the affected function.[39] Today, the question is whether the activation of an area of the brain when a function is being carried out means that this part of the brain "is home to" that function, is the center of ability in math, or faith in god . . . the list is long. We have gone from a negative association to a positive one. But current debates would still benefit from studying those of the past.

To determine the function of a structure through the nature of the deficits caused by its destruction is not a method unique to nineteenth-century neurophysiology, nor, moreover, to physiology in general. But it was the preferred method in genetics throughout the twentieth century, and has continued up to the present. It would also give rise to many disputes among geneticists, and would cause errors in interpretation. Has the debate on cerebral localization influenced geneticists? Or are

39. Finger, 1994, chapter 4; Forest, 2014.

we dealing with a convergence of practices, born of similar experimental constraints?

This debate is not purely academic. The existence of functional cerebral localizations justifies targeted surgical interventions in pathological cases. Lobotomy, which was developed by the Portuguese neurologist Egas Moniz (1874–1955), was highly successful in the 1940s. The current rapid growth of psychosurgical approaches to some diseases is clearly linked to the development of new technologies for exploring the structures and functions of the brain.

7

The Nineteenth Century (Part II)

THE THEORY OF EVOLUTION, THE THEORY OF HEREDITY, AND ECOLOGY

The Facts

Lamarck: An Early Version of the Theory of Evolution

Jean-Baptiste Lamarck first studied botany, and is known for this thanks to his 1778 publication *Flore française* (French flora).[1] The aim of Lamarck's work, which did not adopt Linnaean classification, was to help someone walking through nature, by offering a series of binary choices, to identify the plants encountered along the way. Praised by Buffon, the work assured its author's fame.

The French Revolution, in which Lamarck actively participated at the Jardin du roi (now the Jardin des Plantes),[2] enabled him

1. Burkhardt, 1977; Barthélémy-Madaule, 1979; Corsi, 1988, 2005; Corsi et al., 2006.

2. Chappey, 2009.

in 1794 to be appointed to the chair of insects, worms, and microscopic animals, which he quickly renamed the "chair of invertebrates." Lamarck, who up to then had been a botanist—although he had amassed a large collection of shells—now devoted all his efforts to descriptions of these organisms, as well as of their fossil forms. Between 1815 and 1822, he published the seven volumes of his *Histoire naturelle des animaux sans vertèbres* (Natural history of invertebrates), whose scientific importance would never be challenged.

His appointment to that chair, a result of the chaotic history of the French Revolution, would have fortunate consequences. Indeed, it is unlikely that Lamarck would have produced the same body of work if he had received the chair of botany or that of vertebrates. Invertebrate organisms, with their simple forms, lent themselves well to the transformist scenarios from which he developed the basis of his theory of evolution, much better than plants or mammals would have done.

Lamarck made public his hypothesis of a progressive transformation of organisms for the first time on 11 May 1800, during the opening lecture of his teaching course, and developed these ideas in his *Philosophie zoologique* (*Zoological Philosophy*), first published in 1809.[3]

He built his transformist views on concepts in chemistry, but was strongly criticized for his adherence to the four elements of Greek science, and for his rejection of the chemical revolution of Lavoisier. Convinced of the role that organisms played in the formation of rocks on the Earth's surface, Lamarck planned to write a complete history of the Earth in three parts: hydrogeology, meteorology, and biology. It was while preparing this work that he became aware of the need to add to the role of organ-

3. Lamarck, 1984.

isms in the formation of rocks a parallel section on the role of the mineral world in the genesis and transformation of life, thereby including organisms in a cycle comparable to those existing in hydrogeology and meteorology. In this way, Lamarck incorporated transformations of the living world into those of the inanimate world.

Like most of his contemporaries, Lamarck readily accepted the existence of spontaneous generation. But he added a description of the physical processes that enabled primitive forms to progress through stages up to that of the human being. According to his theory, there are multiple fluids involved in this process, including electrical and magnetic fluids. Although they arrive externally, their action is transformed by their "intussusception" into the organism (a term already used by Bourguet and Buffon), in which they contribute to the formation first of simple organs (such as the digestive tract), then of increasingly complex structures. This is (to use a modern term) a slow process of "complexification."

To this regular process, carried out through physicochemical laws, are added the circumstances in which organisms evolve, that can modify their forms and, to a lesser extent, their internal structures. Circumstances create needs to which organisms respond by changes in their habits. These changes cause transformations in certain parts of the animal—the necks of giraffes become longer, the feet of ducks become webbed—which, transmitted and amplified over generations, enable organisms to adapt to their environment, and lead to the animal forms we know today. These external circumstances generate irregular transformations of organisms, different from the regular transformations that result from the process of complexification; this creates a dispersion of organisms around an ideal "ladder" of beings. In the case of plants, Lamarck acknowledged that the external environment could act directly on them and transform them. In animals, however, he saw it as use and nonuse that were responsible for the development or diminution of organs.

Contrary to how Lamarck's theory is often portrayed, it is neither the desire nor the environment of an animal that causes its transformation, but "interiorized" physical forces, and the animal's changes of habit in response to circumstances.

The model Lamarck proposed has been compared to a set of escalators: new organisms appear continually and undertake the long process of transformation. The model of evolution Lamarck proposed was transformism without history: there is not one history of life, rather an indefinite ensemble of more or less similar histories that repeat themselves.

For a long time, the influence of Lamarck's transformist view was considered to be minimal. The presentation he gave of his model was considered far from completely coherent. It was said that his *Philosophie zoologique* wasn't widely read, and that he had paid the price for his revolutionary involvement—he was even criticized for attempting, like the revolutionaries, to introduce a new word by proposing the term "biology." Cuvier challenged Lamarck's ideas, and presented a caricature of them in his eulogy. It was also claimed that Lamarck died poor and abandoned by everyone.

In fact, Lamarck's lectures were attended by a large audience who came from all over Europe, scientists who helped disseminate his ideas. Nor was Lamarck the only one in France to promote transformist ideas. An anecdotal event might illustrate the influence his ideas actually had. In 1826–1827, the giraffe Zarafa, gifted to the king of France by the viceroy of Egypt, was escorted from Marseille by Geoffroy Saint-Hilaire.[4] Its voyage

4. Allin, 1998; Lagueux, 2003.

and arrival in Paris were huge media events, whereas the arrival of a giraffe at London Zoo in the same year gained little attention. How can one explain the difference, except by noting that through the work of Lamarck the giraffe continued to be associated with the idea of transformation of species?

Jean-Baptiste Bory de Saint-Vincent (1778–1846) was one of the French naturalists who were proponents of Lamarck's ideas. Participating in the voyage around the world undertaken between 1800 and 1804 by Nicolas Baudin, he explored the geography and described the animal and plant populations on Île de la Réunion. A specialist in "microscopic animals," he was convinced of the reality of spontaneous generation and believed in the existence of several human species.

The most famous naturalist influenced by Lamarck's ideas was Geoffroy Saint-Hilaire. At the age of 21, he was appointed to the Muséum d'histoire naturelle, which had been recently created in the Jardin du roi. He participated in Napoleon's Egypt campaign of 1798–1801. In 1818 he began publishing his major work, *Philosophie anatomique* (Anatomic philosophy). Influenced by the ideas in *Naturphilosophie*, Geoffroy Saint-Hilaire was convinced that all animals are built on a single plan: that, philosophically speaking, there is only one animal and there are no new organs—all are just modifications of previously existing ones. He had already expressed this idea in 1795 in his first works on the history of the lemurs of Madagascar, and would present it in its final form in 1822. According to Geoffroy Saint-Hilaire, organisms are formed from the same elements, with the same associations. Influenced by Goethe, he proposed a law of compensation—nature's "budget," as he said, being "fixed": if one part of an organism takes on greater importance, another one diminishes, but does not, however, disappear. Robert Grant (1793–1874), a professor of anatomy at the University of

London, offered him an excellent example of this unicity of plan by demonstrating the existence of a pancreas in mollusks. This added to the homology Geoffroy Saint-Hilaire had demonstrated between the ossicles of the inner ear and the bones present in the gills of fish. The search for homologies—that is, those common elements that are preserved despite their changes in form and possibly in position in an organism—became the number one task of the anatomist. It was not, however, always as fruitful as in the case of the pancreas or gills. Also under the influence of *Naturphilosophie,* the writings of Oken and Goethe made vertebrae an essential element of the anatomy of organisms, and Geoffroy Saint-Hilaire believed he had found an exteriorized homologue of them in insects in the form of traces in the cuticula.

The question of the transformation of species is not central in Geoffroy Saint-Hilaire's thinking. But the existence of homologies suggested a "metamorphosis" of organisms—another term borrowed from *Naturphilosophie*—which, unlike Lamarck, he attributed to the direct influence of the environment. For example, he suggested that lungs formed owing to a shortage of oxygen. Although such important changes were no longer observed in nature, Geoffroy Saint-Hilaire thought that they could have occurred in the past.

By seeking to demonstrate the unity of organization among mollusks and vertebrates, Geoffroy Saint-Hilaire came up against Cuvier, who saw him as an imitator of Lamarck. For Cuvier, the animal kingdom was divided into four branches—jointed animals, vertebrates, mollusks, and radials—each of which possessed an organizational plan unique to that branch. The debate was launched on 28 February 1830, and continued until Cuvier's death in 1832.[5] There were ideological and politi-

5. Piveteau, 1950.

cal ramifications that went well beyond the question being debated. The "morphological" view of Geoffroy Saint-Hilaire, stressing the form of organisms and their parts and inherited from *Naturphilosophie*, clashed with the more physiological and functional theories of Cuvier. Regardless of their personal opinions on these questions, for most of the scientists who attended or were aware of the debate, precision and scientific rigor were on the side of Cuvier.

Geoffroy Saint-Hilaire was also known for his work in teratology. He specifically studied malformations in the human nervous system. He explained their characteristics as resulting from the negative effects of chance lesions that appear during development, combined with the action of the law of compensatory organs described above. The latter, when an organ doesn't achieve its normal size and shape, causes a compensatory growth of other organs, whose effects may be positive or negative.

The Contribution of Georges Cuvier

Georges Cuvier (1769–1832) has a bad reputation.[6] Scientifically, he was a fixist. Unlike Lamarck, who was affected by the Restoration, he was able to pass through the successive political regimes in France with relative good fortune. Unreliable in friendship, and taking advantage of his position and his prestige, he ridiculed the ideas of Lamarck and Geoffroy Saint-Hilaire. He blended his scientific work and his personal religious convictions by identifying the last "revolution of the surface of the globe"—a catastrophic natural event that had wiped out many species—with the flood as described in the

6. Coleman, 1984; Taquet, 2006.

Bible. In 1817, he dissected the cadaver of Sarah Baartman (called the "Hottentot Venus"), a South African woman put on display in European countries because of the size and shape of her buttocks. His insistence on the similarity between her physical characteristics and those of an ape adds an additional dark note to a negative portrait of Cuvier (although he was not alone in these offensive ideas—Geoffroy Saint-Hilaire, examining her while she was still alive a few months earlier, arrived at the same conclusions.[7])

Our objective, following in the footsteps of the historian of science Martin Rudwick,[8] is to amend that caricatural image of Cuvier: not only because he created a remarkable body of work as an anatomist and paleontologist—which is sometimes acknowledged—but above all because he was the first to describe a history of life, which before him neither the fanciful stories of Lucretius or of Benoît de Maillet nor the models of Lamarck and Geoffroy Saint-Hilaire had done.

Born in 1769 in Montbéliard, which at that time belonged to the Duchy of Württemberg, Cuvier studied in Stuttgart. From 1788 to 1794, he was a preceptor in Normandy, which enabled him to carry out his first observations as a naturalist, while escaping the turmoil of the Revolution.

Invited to the Muséum d'histoire naturelle by Geoffroy Saint-Hilaire, he was named acting chair of comparative anatomy. There he continued the work of Félix Vicq-d'Azyr (1748–1794), who had made considerable advances in the science of brain anatomy.[9] In his *Leçons d'anatomie comparée* (Lessons in comparative anatomy), published between 1800 and 1805, Cu-

7. Blanckaert, 2013.

8. Rudwick, 2005, 2014.

9. Schmitt, 2006.

vier noted two essential principles that should guide the work of an anatomist: the principle of the correlation of parts, and that of the subordination of characters. The first asserts that the forms of the various parts of an organism depend on one another because of the functions that they must perform in common—for example, a herbivore organism will have teeth capable of chewing grass, an organization of the digestive tract enabling it to digest it, horned feet to be able to graze, and so forth. For Cuvier, faithful to the view of the organism as expressed by Kant in *Critique of Judgment,*[10] an organism is an assemblage, each part working for the entire organism, and for the other parts. The second principle, borrowed from botany and already used by Linnaeus, asserts that, among physiological systems, some have greater importance than others: the nervous system is more important than the respiratory system, which in turn is more important than other subaltern systems. This distinction enabled the classification of organisms: to determine the order and the family, the organization of superior systems had to be taken into account, while the description of subaltern systems enabled a distinction between genera and species.

Those two principles also guided the work of characterization and classification of the fossils of quadrupeds from the Paris Basin that Cuvier undertook with Alexandre Brongniart (1770–1847). The results were published beginning in 1812 in four volumes under the title *Recherches sur les ossements fossiles de quadrupèdes, où l'on rétablit les caractères de plusieurs espèces d'animaux que les révolutions du globe paraissent avoir détruites* (Research on quadruped bone fossils, in which we reestablish the characteristics of several species of animals that global

10. Kant, 2007.

revolutions appear to have destroyed). Since antiquity, fossils had been considered to be the remains of animals that had once lived on the Earth's surface.[11] But the question of the origin of fossils that didn't resemble any known living creature had elicited many theories. They might correspond to living organisms in parts of the world that had not yet been explored. They could be, as Lamarck believed, ancestors of current living organisms; or they might have existed in the past but disappeared from the Earth's surface without leaving any descendants. The choice of quadruped fossils was meant to eliminate the first hypothesis: species of quadrupeds were small in number and well known, and naturalists thought that there remained too few unexplored lands for their numbers to increase very much. However, the difficulty posed by the study of these fossils consisted in that they were rarely wholly intact: at best, a paleontologist found only a few bones.

The precision of Cuvier's anatomic observations and his use of the law of the correlation of parts enabled him to overcome that difficulty. Legend has it that, upon seeing a single bone, Cuvier was able to reconstruct a complete skeleton. More realistically, but no less convincing for his contemporaries, Cuvier could predict the characteristics of a bone missing from a skeleton—a jawbone, a limb bone—which later fossil discoveries often confirmed.

The originality of Cuvier and Brongniart's work was in combining a description of fossils with one of the geologic layers in which they were found: these simultaneous observations reinforced the identification of fossils and that of geologic layers.

We often remember Cuvier only for his descriptions and identifications of fossils. But his work as a geologist is just as important, if not more so. Simultaneously, and sometimes in collabora-

11. Jordan, 2016.

tion with English geologists (when there was a lull in the Franco-English wars), Cuvier participated in that patient work of describing the history of the Earth. Two results of that work are of particular importance. The first is that the duration of that history was considerably lengthened—although that "revolution" was silent and didn't gain a great deal of attention. This provided the evolution of living forms the time it required to carry out its task. The second is the revelation that the history of the Earth is complex and cannot be reduced to simple schemas such as the gradual emergence of land. Cuvier and Brongniart discovered that there is a successive layering of sediment containing both fresh- and saltwater animals, indicating that on several occasions the sea had re-covered land that had already emerged from it.

Cuvier's descriptions showed that fossil species were different from all known living species. This was a new result, the fruit of the precision of his anatomic descriptions. When the German anatomist Peter Simon Pallas (1741–1811) a few years earlier had discovered fossils of mammoths during his exploration of Siberia, he had not been able to see that they represented a new species, distinct from contemporary elephants.[12] Furthermore, each fossil species appeared only in one or a few geologic layers: Cuvier's observations showed that different living forms succeeded one another during the history of the Earth, the most complex forms appearing only in the most recent layers.

Cuvier's anatomic observations, and the law of the correlation of parts that he deduced from them, convinced him of the fixity of species. By comparing the ibises that the ancient Egyptians embalmed and those present in museums—in particular, those stolen by French soldiers in Holland—Cuvier showed they were perfectly identical. And, according to Cuvier, since one could not

12. Cohen, 1994.

confirm what a great span of time would produce except by extrapolating from what is produced over a smaller amount of time, he found an additional argument in favor of the fixity of species.

The interpretation of the entirety of his observations that Cuvier presented in 1822 in his *Discours sur les révolutions de la surface du globe, et sur les changements qu'elles ont produits dans le règne animal* (Discourse on the revolutions on the surface of the globe, and on the changes they have produced in the animal kingdom) is that different organisms have successively populated the surface of the Earth.[13] Since intermediary fossil forms had not been found—yet another argument in favor of the fixity of species—it must be imagined that over time existing forms of life were eliminated and replaced by new ones. To use "revolutions" to describe such events is not really surprising coming from someone who, at the age of 20, even from afar, lived through the French Revolution. Cuvier isn't explicit about these replacements. He doesn't talk about creation, and it isn't unimaginable that all living species had existed since the most ancient of times. What Cuvier was getting at in his reference to the book of Genesis was not to provide support for his observations with revealed truths—that would be a complete misinterpretation of his approach.[14] For him, the Bible was a tale of events as reported by human beings. The Genesis story is but one of the very many tales from human populations that Cuvier presents over more than 100 pages, all of which describe the occurrence of a cataclysm in ancient times. Apart from that reference to the flood, the story of the history of the Earth that Cuvier offers has very little to do with the Biblical text. Cuvier was a positivist. None of the facts that he observed, regarding either living animals or

13. Cuvier, 1985.

14. Buffetaut, 2002.

fossils, convinced him of a transformation of living species. The often ill-informed speculations of Geoffroy Saint-Hilaire, for Cuvier, could not make up for an absence of facts.

Cuvier's goal was to make the archives of the Earth talk, just like the archaeologists who, at the end of the eighteenth century, brought the cities of Pompeii and Herculaneum back out of the ashes under which they had been buried. Just as Buffon had attempted to do before him with a history of the Earth, Cuvier brought to light a history of life.

It might seem paradoxical to talk about a history of life without including the transformation of living species. A comparison with Lamarck is enlightening, because Lamarck did the opposite: he suggested that living species were transformed, though without producing a history of life. Evolution, in the sense that we give the term today, is the sum of transformations and a history. Lamarck and Cuvier each supplied one of the two elements: it was Darwin who brought them together.

Alcide Dessalines d'Orbigny (1802–1857) carried on Cuvier's ideas. A naturalist specializing in foraminifers, he explored South America between 1826 and 1833, and occupied the chair in paleontology created especially for him at the Muséum d'histoire naturelle. He is the author of a monumental work entitled *Paléontologie française* (French paleontology).[15]

The Second Wave of Transformism: Darwin

Charles Darwin (1809–1882) went to Edinburgh to study medicine when he was 16.[16] There he met Robert Grant, who had collaborated with Geoffroy Saint-Hilaire, and he published a

15. Orbigny, 1860.

16. For more about Darwin's life, see Browne (2010a, b).

scientific article with him. Uninspired by medical school, he went to the University of Cambridge to study theology. He devoted a large part of his three years at the university to exploring the world of botany with John Henslow (1796–1861). That encounter would be decisive: Henslow invited Darwin to embark on the *Beagle* for an exploratory expedition around the world. From 1831 to 1836 Darwin sailed along the eastern and western coasts of South America, crossed the Pacific, and reached Australia. During the voyage he read *Principles of Geology* by Charles Lyell (1797–1875), and recorded his observations on the formation of mountains in South America, on islands, and atolls. He collected plants and animals, as well as a large number of fossil bones. He also observed indigenous populations—the inhabitants of Tierra del Fuego, and Australian aborigines.

Thanks to the observations that he sent back home during his trip, his scientific reputation was already well established when he returned to England. His first published works were in geology. The account of his voyage, initially incorporated into the official report of Captain Robert FitzRoy, was later published as a separate work. It was hugely successful, and made Darwin known to a wide audience.[17] He passed around the samples he had collected to have them analyzed, while keeping some, which he studied himself. He gave the fossil bones he had brought back from South America to the famous paleontologist Richard Owen. Occasionally, Darwin's errors were thus corrected: what he thought to be varieties of finches that he had brought back from the Galapagos Islands turned out to be different species.

During the expedition, Darwin had already begun thinking that the idea of a separate creation of the different species exist-

17. Darwin, 2001.

ing on the Earth's surface was false; the only way to explain the similarity between the fossils discovered on one part of the globe and the organisms that currently lived on it, or the presence on islands of species that were similar but not identical to those found on the closest continent, was to admit a progressive transformation of species. Just one year after he returned, Darwin sketched in his notebook an "evolutionary tree" showing how species could diverge from one another. The following year, in 1838, he found the mechanism for these transformations while reading a reprint of Malthus's work *An Essay on the Principle of Population*. In it, Malthus explained conflicts among humans by a discrepancy between a slow increase in food resources—an arithmetical increase—and a rapid geometrical growth (today, we would say exponential) of the population. Darwin transposed this schema to the animal and plant worlds: the growth of populations leads to competition for food resources in which only the best-adapted varieties survive. He wrote an initial version of his theory in 1842, and a more developed one in 1844.

Over the following years, Darwin accumulated observations that supported his theory by using two strategies. The first was to collect precise information, thanks to a network of correspondents, botanists, and zoologists. He also questioned breeders, convinced that the transformation of species occurred by a process of selection analogous to that used specifically by breeders of pigeons. Darwin's second strategy was to collect his own data. He described and classified the different species of Cirripedia, or barnacles, studied the mechanisms of fertilization of orchids and competition among plants in his own garden, and experimented with the transportation of seeds and eggs—with the aim of understanding how islands are populated.

In 1858, he received an article from the naturalist Alfred Russel Wallace (1823–1913) in which Wallace proposed a mechanism for the transformation of species through competition, something also inspired by his reading Malthus. Darwin sent the article to be published in the *Journal of the Linnean Society*, but Charles Lyell and Joseph Hooker (1817–1911) intervened so that the publication wouldn't knock Darwin's discovery out of first place. They added two texts by Darwin to Wallace's article: a montage of excerpts from the 1844 text, and a letter of 1857 addressed to the American naturalist Asa Gray in which Darwin offered a summary of his theory. The three texts appeared together.

In 1859, Darwin published *On the Origin of Species by Means of Natural Selection, or the Preservation of Favoured Races in the Struggle for Life*,[18] which was hugely successful. The 1250 copies of the first printing were all bought through subscription, and there were multiple subsequent reprintings. The *Origin of Species* is, in spite of its size, a summary. Darwin had envisioned a work that would assemble all the observations he had collected in support of his theory. What couldn't be included in the *Origin of Species* would be published in three works: *The Variation of Animals and Plants under Domestication* (1868), *The Descent of Man, and Selection in Relation to Sex* (1871), and *The Expression of the Emotions in Man and Animals* (1872).[19] While the *Descent of Man* also presented the results of his observations on sexual selection, the third work is atypical. The idea of a continuity in the expression of emotions among animals and human beings is more Lamarckian than Darwinian. The use of many drawings and photographs also gives this book a distinctive

18. Darwin, 2009.
19. Darwin, 1981, 1989, 1998.

character. One can see in Darwin's ideas a precursor of ethology; without forgetting, however, that the work is more a projection of human emotions onto animals than an objective study of their behavior.

Darwin also published several works on subjects that weren't directly related to his theory, but contained material in which he often found arguments in its favor: a monograph on the subclass of Cirripedia (1851); *On the Various Contrivances by which British and Foreign Orchids are Fertilised by Insects* (1862); *The Movements and Habits of Climbing Plants* (1865);[20] a work on insectivorous plants (1875); and *The Formation of Vegetable Mould through the Action of Worms* (1881).

Although he had not been convinced of the transmutation of living species until after his return to England, Darwin had very early on been aware of transformist ideas. His grandfather, Erasmus Darwin, had advanced such ideas in his work *Zoonomia* (1794–1796). And as a young man Darwin had worked with Robert Grant, who was a proponent of transformism and was very familiar with the ideas of Geoffroy Saint-Hilaire and Lamarck. Later, Darwin would also have contacts with the group that Robert Grant had assembled around himself.

And so the idea that living species could be transformed wasn't new. It even became popular, as seen in the success of the work by Robert Chambers *Vestiges of the Natural History of Creation*, published anonymously in 1844.[21] Those most reluctant to accept transformism were the naturalists, who were shocked by the fantastical transformations found both in the writings of Lamarck and Geoffroy Saint-Hilaire and in more popular works as well.

20. Darwin, 1988.
21. Secord, 2000.

Darwin was quickly convinced that the only way to have the idea of evolution accepted was to provide precise and incontestable scientific observations to support it. From beyond the grave, Darwin's adversary was Cuvier. Darwin had to provide as many arguments in support of an evolutionist theory as Cuvier had provided to support his fixist concept. He didn't want to repeat the mistakes of Geoffroy Saint-Hilaire and Lamarck. Darwin's insistence that evolution is gradual and proceeds in small steps, which even his disciples found unrealistic, has several explanations. The first, a simple one, is that it is such variations that one observes most often in nature, the variations with which breeders are concerned. Thus, Darwin applied the principle of actualism, dear to the geologist Lyell, to the transformations of organisms: call upon only the causes whose effects are observable today to explain the phenomena of the past. Another reason, as suggested by the paleontologist Stephen Jay Gould, for Darwin to favor gradualism, is that he didn't want to have sudden and great "miraculous" variations be the drivers of evolution. The third is probably more important: only small-scale variations on the level of the organism are likely to be tolerated by living systems, a notion compatible with Cuvier's principle of the correlation of parts.

It was this will to build an irrefutable theory that explains both the time that went by before the publication of the *Origin of Species* and the success of the work. By seeking arguments in all realms of biology, from biogeography—a science founded by Alexander von Humboldt that consists of describing the distribution of living beings on the Earth's surface—to paleontology and embryology, Darwin offered a converging assemblage of arguments supporting the evolution of living beings.

He also provided a mechanism that explained this evolution: the natural selection of advantageous variations. Thus, evolution doesn't follow lines of development, an idea that Lamarck had proposed and Chambers had made popular, but is the result of an adaptation to constantly changing environments. This second aspect of Darwin's work was probably less immediately important—on the one hand because many of his readers would turn it into a secondary mechanism, or a sort of proof according to which ill-adapted organisms would not survive. On the other hand, Darwin himself didn't reject other mechanisms, such as the transformation of habits and the transmission of those changes to descendants. The idea that evolution, which is a historical process, should obey a mechanism was an idea that didn't trouble Darwin's contemporaries, who were accustomed to seeing in human history a process guided by laws. Retrospectively, it seems more ambiguous to us.

For Darwin, the phenomenon of evolution led him to question all existing classification systems, and in particular that of Linnaeus: the only valid classifications were henceforth those based on genealogy.

Darwin's hesitation, his fear of possible negative reactions, and the violence of those reactions have been greatly exaggerated. Darwin did hesitate, but that was in his nature: he might have favored a hypothesis, but he did not eliminate competing hypotheses if he didn't have the arguments to do so. Negative reactions were limited, even in religious circles. Scholars always cite the confrontation between the bishop of Oxford, Samuel Wilberforce, and Thomas Huxley during a session of the British Society for the Advancement of Science. But they forget to mention that in 1864 Darwin received the prestigious Copley Medal from the Royal Society, and that in 1882 he was buried in Westminster Abbey next to Newton.

Natural selection is often presented as one of the rare examples in the history of science where the same theory is proposed simultaneously and independently by two scientists. Actually, the theories of Darwin and Wallace are not identical. Darwin stressed competition among individuals of the same species, whereas Wallace conceived of the possibility of competition among groups, and even among different species. That question would again be debated at length by twentieth-century evolutionary scientists. The second divergence involves the role of sexual selection in evolution. Sexual selection often consists of a choice of males by females at the time of mating. There isn't competition for food, but direct competition in the mating process. Darwin describes this in his 1871 book, and assigns it particular importance in human evolution. Wallace accepted its role only in cases where it would merge with natural selection—when, for example, competition leads to a battle between males. The third divergence between Darwin and Wallace has to do with the role of artificial selection. Darwin considered it a useful analogy, even though he believed that natural selection is more effective because it acts not on appearance, but on internal organs. For Wallace, there was nothing in common between artificial selection and the competition of organisms in nature. The fourth difference involves the inheritance of acquired characteristics. Wallace rejected it completely, whereas Darwin assigned it a place next to the natural selection of random variations.

Another difference appeared in the years following the publication of the *Origin of Species* concerning the subject of human evolution. Whereas Darwin, allusive in his first work, subsequently fully included human evolution within the evolution of living organisms, Wallace did the opposite: he asserted the irreducibility of human evolution to that of other organisms, due to the presence of a spiritual principle in the human being.

Contradictory portrayals of Wallace have appeared.[22] Some depict him as a victim of Darwin, stripped of his discovery by Darwin and his friends. Others describe him as a pseudoscientist, as evidenced by his belief in phenomena of spiritism. The reality was more complex. Wallace was a recognized naturalist who had traveled through Amazonia and the Malaysian archipelago. He was the author of many original observations, in particular of mimicry in the animal and plant worlds. The phenomenon of mimicry was studied at the same time by both the German naturalist Fritz Müller (1821–1897) in Brazil and the British entomologist Henry Walter Bates (1825–1892)—with whom Wallace had carried out his first exploratory expedition in Amazonia. It involves a living species mimicking another animal or plant species, and could be interpreted very well within the framework of the theory of natural selection and, for that reason, provided strong evidence for it. Wallace also published important works on the geographic distribution of animals. He never believed that Darwin had stolen his discovery from him—on the contrary, it was he who gave the theory of evolution the name "Darwinism." He allowed himself to be led astray by the followers of spiritism, as he had not been trained in the experimental method or the scientific rigor it demands as thoroughly as many of the scientists of his time.

The Theory of Heredity

On 8 February and 8 March 1865, in two lectures given at the Natural History Society of Brno, Gregor Mendel (1822–1884) presented the results of experiments he had been conducting for eight years in the garden of his monastery on the hybridization

22. Fishman, 2004; Reisse, 2013.

(crossing) of varieties of peas.[23] From his observations he had derived two laws governing the formation of hybrids: the law of segregation, and the law of independent assortment. The first holds that parental characteristics that join in the hybrid separate during the formation of the hybrid's reproductive cells; the second, that different characteristics are transmitted independently.

Mendel was born in a rural village in Moravia. He was preparing to attend the University of Olomouc, but chose instead to enter the Augustine monastery of Brno, where he was ordained a priest. In 1851 and 1852, he took courses at the University of Vienna. Returning to Brno, he taught physics and natural sciences, began his work on hybridization, took care of the monastery's beehives, and performed meteorological observations. In 1865, he founded the Austrian Meteorological Society, which coordinated the network of meteorological stations established throughout the Austrian Empire. In 1867, he was named abbot of the monastery, a position he held for the rest of his life.

The best way to describe the life and work of Mendel is first to eliminate, one by one, the erroneous versions of them that have circulated. Mendel was not a poor, uneducated monk from a remote region of Europe. Moravia was a rich agricultural region, located not far from the capital, Vienna, where it sold its abundant harvests. Techniques for improving animals and plants were implemented there, and research was actively conducted toward that goal. Mendel, like Darwin, would find in these new agricultural practices—hybridization for him and artificial selection for Darwin—a source of inspiration for his scientific work. The mission of the monastic order that Mendel joined was education. It was an order open to the world: when

23. Pichot, 1999.

Mendel entered it, the monks wore the clothes of laymen. The prior, Cyril Napp, had created a botanical garden there to be able to carry out research on the improvement of plants.[24] Even though Mendel twice failed the recruitment exam for university professors, it is not reasonable to deny that he was a scholar. At the University of Vienna he took courses in mathematics as well as those of the physicist Christian Doppler (1803–1853), and learned of the cellular theory of Rudolf Virchow.

His work on the hybridization of peas is a model of scientific rigor. All his experiments were done through artificial pollination. Mendel took great care with his choice of the seven characteristics whose transmission he studied. He chose them because they were stable and because hybrids had the characteristic of only one of their two parents (the dominant characteristic, as opposed to the other, recessive, characteristic—those terms were coined by Mendel) and because the hybrids were as fertile as the parent plants. He carried out a large number of crosses because he was convinced that only a statistical approach would reveal the laws he was searching for. This, in fact, enabled him to go beyond the observations on hybridization that had already been made by other naturalists, such as the Frenchman Charles Naudin (1815–1899).

The greatest mistake, perhaps, is to see Mendel's life as a series of imposed choices. For example, that Mendel entered the order for economic reasons and to be able to live a simple and studious life—there is no reason to think that his vocation wasn't sincere. Or that Mendel's results would have not been well received, which would have led him to abandon his work—in fact, it seems that his peers received them favorably. It is true that the scientists to whom Mendel sent the texts of his lectures,

24. Wood and Orel, 2001.

such as the great botanist Carl von Nägeli, didn't really see the importance of them. But there is no proof that it was disappointment that pushed Mendel to abandon his research and agree to head the monastery. However, he encountered difficulties reproducing his results on hawkweed (*Hieracium*), as Nägeli had suggested he do, and also on honeybees, which did block the development of his work. Mendel had another scientific passion, meteorology, which he never abandoned. Nor is there evidence that he was unhappy to be named abbot of the monastery. At his funeral, the composer Janacek himself played the music, showing that Mendel was a notable greatly respected by his town.

The final myth concerns his results, which, according to the population geneticist Ronald Fisher, were "arranged" to better reflect the laws Mendel expected. Mendel himself wrote in his 1866 articles that, convinced of the existence of the laws he had discovered, he had ignored the results of certain crosses that were not close enough to them, considering that the experiments that had produced those results must have been badly conducted. Was Mendel's "wrongdoing" more serious than this selection of data? Many articles have been published on that question; they lead to no clear conclusion, mainly because no one knows the precise procedures Mendel used.[25]

The weak impact his results made has other explanations. His work was situated within an area of research that had been going on for many years, aiming to describe the characteristics of hybrids formed by crossing plants or animals.[26] A central question was the stability of those hybrids. Hybridization is "the" technique used to create new varieties combining the

25. Hartl and Fairbanks, 2007; Franklin et al., 2008.

26. Brannigan, 1979; Olby, 1979.

qualities of two parents, but the instability of hybrids is an obstacle to the accomplishment of such projects. Mendel confirmed that if, from the first generation of hybrids, one systematically crosses their own descendants, the percentage of hybrids regularly decreases in successive generations.

Furthermore, Mendel's results applied to characteristics. Even if, in his article, Mendel had sought a mechanism that was likely to explain the results he observed, he was never able to clearly conceive of the existence of the element—we would call it a gene—that carried the characteristic without itself being the characteristic. Mendel introduced modern notation: uppercase for the dominant allele and lowercase for the recessive allele. The hybrid is always designated by a single letter, uppercase, showing that the description and the writing focus on the characteristic, and nothing else.

The use of mathematics probably also put off a number of naturalists. Writing out the results of the crossing of peas with three different types of characteristics necessitated five lines of symbols, something completely unheard of in biology, even if it was only the (simple) development of a polynomial.

This in no way detracts from Mendel's contribution: to have used for his study well-defined characteristics and achieved enough crosses to deduce convincing regularities from them, in spite of the random combination of sex cells bearing one or the other form of the characteristics.

But, in 1865, a theory of heredity had not yet been developed. There were two competing events: the genesis of a model—false—by Darwin, and the observations by cytologists on cell division and fertilization, with the identification of chromosomes and the speculation that created.

Darwin proposed his theory of heredity (the theory of pangenesis) in the final chapter of his 1868 work *The Variation of*

Animals and Plants under Domestication.[27] He was aware that a description of the mechanisms by which variations are transmitted from generation to generation, to which he attributed the progressive evolution of species, was an indispensable complement to the theory of evolution. He first presents the facts that his hypothesis would explain, he summarizes the hypothesis in two pages, and finally shows that the hypothesis is compatible with observations. Darwin's ambition in the realm of heredity was as strong as in that of evolution. He wanted his hypothesis to explain all the phenomena of heredity known in the animal and plant worlds, including the effect of transplants, regeneration, the inheritance of acquired characteristics, the facts of atavism—the resurgence of ancestral traits—and even phenomena of impregnation, that the sexual relations of a female with a male might modify the characteristics of her later descendants.

The theory Darwin was proposing hardly differed from that of Hippocrates and Buffon—as he would point out, moreover, in later editions of his work: Particles, which he calls gemmules, emitted by different parts of the organism, are found in sex cells. After fertilization, they participate in the construction of the organism.

Such a model fully justified the inheritance of acquired characteristics: an organ that has increased or grown stronger during the life of an organism will send more gemmules to the reproductive organs and the descendants will then reproduce the characteristics of their parents. This led Darwin in the later editions of the *Origin of Species* to assign increasing importance to the inheritance of acquired characteristics. Gemmules could be active or dormant, which would explain phenomena of atavism.

27. Darwin, 1998.

Francis Galton (1822–1911), Darwin's cousin, tested this hypothesis from 1868 to 1871 by transfusing rabbits with the blood of rabbits of different species. He showed that the descendants of the transfused rabbits showed none of the characteristics of the rabbits whose blood they had received, which, in his opinion, invalidated Darwin's theory. Darwin protested, but recognized that his theory wasn't precise enough on the mode of transmission of gemmules. Furthermore, it was not easily compatible with cellular theory: not only did gemmules circulate from cell to cell, but, according to Darwin, they would be capable of combining to form new cells. This led Darwin to suggest that cellular theory was far from being firmly established, echoing a criticism that Thomas Huxley had voiced a few years earlier.[28]

The theory of pangenesis was a complete failure: it wasn't original, clashed with the experimental data, and was incompatible with other theories in biology. Darwin would experience similar disappointments in the work on plant biology that he carried out with his son Francis: they would be called "amateurs" by the great German physiologist Julius Sachs.[29]

However, the theory of pangenesis was a failure with fortunate consequences: Darwin had opened a path that other biologists would follow by proposing models more compatible with the observations of cytologists. They would, according to the terms of the historian Garland Allen, gradually develop a materialist and corpuscular theory of heredity.[30]

The first to do this was August Weismann (1834–1914), a professor at the University of Freiburg. Afflicted from a very young

28. Richmond, 2000.

29. Bernier, 2013.

30. Allen, 1975.

age with a disease of the eyes, he abandoned his microscope work and turned to theory. In 1883, in a lecture titled "On Heredity," he rejected any form of inheritance of acquired characteristics. For him, germplasm—a notion related to that of idioplasm proposed earlier by Nägeli—was the material support of hereditary phenomena. It was present in the nucleus of cells of the germline. The germline experiences a unique development during embryogenesis, distinct from that of the other parts of the organism, or the soma. Even if the germplasm were affected by modifications that affect the soma, it would not be able to reliably transmit them to the organism's descendants. Weismann pointed out, moreover, that none of the facts of so-called inheritance of acquired characteristics had been clearly established; and though it is impossible to deny a priori the existence of such phenomena, it is simpler to dispense with them and to make the sorting of variations through natural selection the only mechanism of evolution. The change Weismann thus made to Darwin's theory of evolution was so important that one henceforth spoke of "neo-Darwinism" to designate Darwin's theory rid of the inheritance of acquired characteristics.

In 1892, in his work *Germ-Plasm, a Theory of Heredity,* Weismann honed his concept of heredity, relying on the recently acquired knowledge of mechanisms of fertilization and the role of chromosomes. He developed a complex hierarchical model: biophores are elemental units of life capable of reproducing; several biophores join to form a determinant; determinants in turn combine into ids, and ids into idants, which Weismann cautiously identified with chromosomes. He added a functional element to this hereditary model: biophores reach the cytoplasm, where they participate in the formation of characteristics. By eliminating the inheritance of acquired characteristics, Weismann had deprived himself of a simple explanation of the origin of variation. He attributed variation to the recombination that occurs during sexual reproduction—but this wasn't a true creation of something new. He also thought that among organisms that have no separation between the soma and the germ cells, or germen, biophores, subjected to the action of the environment, would be able to diversify, and that this new variation would persist in organisms that did have separation between soma and germen, as it had been generated before that separation. Finally, he called upon a mechanism of competition among biophores, borrowed from Wilhelm Roux (1850–1924) and his idea of competition among cells during embryogenesis.

In contrast, the question of the origin of variation was central to the work of the Dutch botanist Hugo de Vries (1848–1935). He spent several years in Germany, first in Heidelberg, with Wilhelm Hofmeister (1824–1877), the specialist in physiological chemistry. Hofmeister revealed the importance of what was called the biggest discovery in plant biology of the nineteenth century: the alternation of generations. The life cycle of a plant includes two phases, an asexual sporophyte phase and a sexual gametophyte phase. Sporophytes and gametophytes can be independent or combined multicellular structures, of equivalent size, or not. In flowering plants, the gametophyte is reduced to a few cells within the sporophyte. The significance of these two phases was not fully understood until after the discovery of chromosomes, and the demonstration that sporophytes are diploid, containing two sets of chromosomes, whereas gametophytes are haploid, containing only one set. De Vries later worked in Wurzburg with the plant physiologist Julius Sachs, and was named professor in Amsterdam in 1878.[31]

31. Vallade, 2008.

In 1889 he published a work titled *Intracellular Pangenesis*. The model he develops in it is close to the one that August Weismann would propose much more precisely three years later. The pangenes de Vries discusses, the equivalent of biophores and like them capable of reproducing, are the origin of the term "genes." In 1900 de Vries would be, as we will see, one of three scientists who rediscovered Mendel's laws. Shortly after this rediscovery, he published a work titled *The Mutation Theory*, the result of research begun in 1886 on *Oenothera lamarckiana*. In 1889, he described in the book the spontaneous appearance of large variations that he called "mutations," occurring at a frequency of around 1 in 1000, impacting various characteristics of the plant, and reliably transmitted to descendants. For De Vries, these mutations were the essence of evolution. He would try in vain to reconcile the existence of mutations with the model of pangenesis that he had published a few years earlier.[32]

The theory of evolution had been shaken. Not only is the germplasm perfectly stable between two mutations, and the characteristics of the organism constant, but the evolutive novelty is in the mutation that, in a single step, creates a new species. Natural selection still intervenes, but it no longer involves a competition between individuals of the same species, but between old and new species. For some biologists, the hypothesis of mutations reconnected with the catastrophist tradition of Cuvier.

The irony of history is that the mutations observed by de Vries were not mutations in the sense that geneticists would give the term in the first half of the twentieth century. The mutations observed by de Vries in *Oenothera lamarckiana* were the fruit of complex phenomena of transposition.

32. Stamhuis et al., 1999.

In contrast to that work, Francis Galton developed an approach to the phenomena of inheritance that was completely different.[33] He was a precocious genius, an explorer in Africa, inventor of weather maps—and the first to have attempted to quantitatively measure the effectiveness of prayer. He proposed a method for classifying fingerprints and estimated the probability of finding two identical prints. His results led to their use in legal procedures. He is also considered the father of eugenics (1883).

His cousin Charles's book had an important impact on Galton. He began to look into the hereditary transmission of characteristics, both physical and mental, and in 1869 he showed in his book *Hereditary Genius* that the "innate" is more important than the "acquired." He also completely rejected the existence of the inheritance of acquired characteristics. Most of Galton's work deals with measurable characteristics, such as height and weight. He studied their distribution in a population, characterized by an average and a variance. He observed their transmission over generations and noticed that a regression toward the average was very commonly seen. His work was continued by the statistician Karl Pearson (1857–1936) and the biologist Raphael Weldon (1860–1906), who in 1901 founded the journal *Biometrika*. The approach of biometricians can be distinguished in two ways from the work described above, and from that of the geneticists who would come after. Biometricians were interested in "measurable" characteristics, whereas Mendel, like Darwin, Weismann, and De Vries, studied above all qualitative characteristics (color, shape). They sought laws, and weren't interested in the mechanisms that might be at their origin. In that, they were followers of the empirical and positivist traditions in science of August Comte and Ernst Mach (1838–1916).

33. Gillham, 2001.

The Reception of Darwin's Theory and the Eclipse of Darwinism

The reception of the *Origin of Species* varied greatly from country to country, and by scientific discipline: most paleontologists were against it. Even biologists who were largely favorable did not always agree with all of Darwin's ideas, which they often distorted, either consciously or unconsciously.

The theory of evolution was well received in Great Britain. Darwin benefited from active and consistent support, such as that of Joseph Hooker and Thomas Huxley (1825–1895). Hooker was a field botanist who participated in many expeditions to Antarctica and the Himalayas and in the western United States, before becoming director of the Royal Botanic Gardens in London. Darwin gave him some of the samples he had brought back from his voyage on the *Beagle*, and informed Hooker of his new theory very early on. Hooker helped prepare the joint publication with Wallace in 1858, was a firm supporter of the *Origin of Species*, and, by his own claim, participated successfully in the famous debate of 1860 with Bishop Wilberforce.[34]

Huxley is considered to have been Darwin's most avid supporter, his "bulldog." However, Huxley came to subscribe to the idea of evolution only belatedly, for a long time remained reticent vis-à-vis gradualism, and accepted the role of natural selection only for lack of a better explanation.[35] Like Darwin, he had undertaken an expedition lasting several years—on the *Rattlesnake*—during which he studied marine invertebrates, in particular jellyfish. At the end of that voyage, he was appointed chair of natural history at the Royal School of Mines. Huxley

34. Endersby, 2008.

35. Lyons, 1999.

participated actively in the debates that followed the publication of the *Origin of Species,* and in his 1863 work *Evidence as to Man's Place in Nature,* he presents a new view of the human species as provided by the theory of evolution. During a famous dispute with the paleontologist Richard Owen (1804–1892), he demonstrated that, contrary to what Owen had claimed, the brain of human beings was not distinguishable from that of apes by the possession of specific structures.

Huxley was the perfect representative of those many biologists who promoted the theory of evolution as a weapon for the advancement of science and a reform of education, against all the conservative, religious, and political forces of the time. His student Ray Lankester (1847–1929), professor at University College London and director of the British Museum of Natural History, continued Huxley's scientific and political battles—and was a friend of Karl Marx. Adhering to the ideas of Weismann, and thus convinced of the major role of natural selection in evolution, he specifically studied phenomena of degenerescence in invertebrates (explainable by an absence of the action of natural selection necessary for the maintenance of an organism's characteristics). He adopted the method proposed by Ernst Haeckel to establish animal phylogenies through a comparison of the first stages of embryonic development (see below), a realm in which he made many original observations.

The geologist Charles Lyell also supported Darwin. Doubting (as early as in 1830) the stability of living forms, he remained influenced by Lamarck's ideas and was reluctant to give natural selection a major role. For religious reasons, he always set the human being apart.

One of Darwin's most intractable adversaries, as we have seen, was the paleontologist Richard Owen. From the volume of his work as an anatomist and paleontologist, but also from the position he held and his authoritarianism, he can be compared to Cuvier.[36] But "impregnated" with *Naturphilosophie*, his ideas were closer to those of Geoffroy Saint-Hilaire than to Cuvier's. He is famous for his reconstruction of the skeletons of dinosaurs, a term he invented. Under his direction, reconstructions of prehistoric animals were exhibited at the Crystal Palace in London, much to the delight of the public. Before the publication of Darwin's *Origin of Species*, Owen had accepted the idea of a limited transformation of species based on archetypes, a transposition of Baer's embryologic model to the evolution of organisms.[37] But he was adamantly opposed to Darwin and his disciples, refusing to accept that natural selection was the driver of evolution: for him, evolution was internal to organisms, not external. Mechanisms such as those leading to the repetition of parts, he thought, would be responsible for the transformation of organisms. Owen made a lot of enemies by claiming credit for results that weren't his. In particular, he didn't attribute to Gideon Mantell (1790–1852) the discovery of the iguanodon, a giant dinosaur. For that reason, and because of his failure with Huxley, he was not one of Darwinism's strongest adversaries.

After reviewing Darwin's book favorably, the English naturalist Saint George Jackson Mivart (1827–1900) in his 1871 work *On the Genesis of Species* revealed the limits of Darwinian theory. Not only had the study of fossils up to then not revealed the intermediary steps that Darwinian evolution assumed to exist, but the mechanism of natural selection was unable to explain either the appearance of new organs or the evidence of parallel evolution. Mivart thought that there were large-scale instantaneous changes in evolution. Darwin's theory, then,

36. Schmitt, 2006.
37. Camardi, 2001.

needed to be extended, since these variations obeyed laws. Interestingly, Mivart was advancing an argument that would be found among the French neo-Lamarckians, even though his (religious) motivations were very far from those of the latter. Darwin attempted to respond to Mivart's objections, but the men nonetheless ended up falling out.

William Carpenter (1813–1885), an English physician and zoologist and a specialist in the nervous system of invertebrates, lent his support to Darwin, but also had reservations concerning the application of the theory of evolution to the intellectual and spiritual nature of the human being.

In the United States, reactions were also very mixed. Asa Gray (1810–1888), a botanist, specialist in American plants, and professor of natural history at Harvard University, was a very close friend of Darwin and Hooker. His strong religious convictions in no way prevented him from accepting the theory of evolution through natural selection that Darwin proposed in the *Origin of Species*, and he was instrumental in having the book published in the United States.

The reception of the book by American paleontologists was less enthusiastic. Louis Agassiz (1807–1873), a student of Humboldt and Cuvier, was a specialist in the anatomy of living and fossil fish.[38] He also confirmed earlier observations on the role of the last great ice age in the morphogenesis of landscapes. A professor at Neuchâtel, he became director of the Museum of Comparative Anatomy at Harvard. He rejected the Darwinian theory because his observations as a paleontologist did not seem to align with the progressive view of evolution promoted by Huxley, and even more so by Haeckel (see below). The variations in organisms that one could observe in nature did not

38. Morris, 1997.

allow the species barrier to be crossed. He believed, as did Cuvier, that the creation of new species was stimulated by periodic catastrophic events. He rejected all transformist ideas, yet throughout his life as a scientist insisted on the triple parallelism of series in systematics, embryology, and paleontology. He was a proponent of polygenism—that is, of the idea of a separate creation of the different human races—although he was convinced of the equality of the races on a spiritual level. The American physician and naturalist Samuel George Morton (1799–1851) shared the same polygenist view. However, in his opinion the collection of skulls that he had assembled in Philadelphia showed differences in the size of the brain among the different human races, in particular between White and Black people.

More interesting from the point of view of biology was the attitude of the American paleontologist Edward Cope (1840–1897).[39] Along with Othniel Marsh (1831–1899), he was one of the two great fossil prospectors in the American West, working during the Western expansion and the building of the railroads. Their rivalry in the years 1870–1880 would lead to what has been called the "bone wars." Cope, unlike his adversary Marsh, accepted the idea of evolution, but not the mechanism Darwin proposed. He thought that the action of natural selection could explain the formation of secondary traits, but not the transformations that a paleontologist observes over longer periods. Those changes, in his opinion, were due to the acceleration of embryonic growth, which enabled the addition of supplementary stages of morphogenesis. Embryonic development could also slow down, which led to degenerations and extinctions. Cope's ideas evolved gradually; he rejected the idea that there

39. Bowler, 1977.

was a single mechanism of transformation, leaned toward Lamarckian concepts, and thus became one of the founders of American neo-Lamarckism. Use and nonuse he thought played a major role in the development of organisms. There was an energy that Cope equated with intelligence and consciousness, which, transmitted from generation to generation, grew some parts of the organism to the detriment of others. Cope left his name on several "laws" of paleontology that were successful for a limited time: a law according to which species had a tendency to become larger in the course of evolution, and one where primitive species evolved more quickly than more evolved species.

Darwinism had a spokesman in the Germanic world, as it had in Great Britain with Huxley, in the person of Ernst Haeckel (1834–1919); but he was a much less loyal spokesman. Haeckel was a professor of comparative anatomy at the University of Jena.[40] He worked on radiolars, large single-celled organisms with an external skeleton of silica, present in zooplankton. He also studied sponges and jellyfish. He introduced the term "Monera" to describe all organisms without a nucleus. Haeckel was a great inventor of new words, only some of which (such as ecology) would survive. He was also a brilliant illustrator, and his drawings would contribute greatly to his renown.[41]

He read Darwin in 1864 and was immediately attracted by his ideas. He extended Darwin's work in two directions: human evolution, for which he anticipated the discovery of an intermediary between apes and human beings, which he called the "Pithecanthrope," and the representation of evolution in the form of trees. Where Darwin had only sketched an abstract

40. Nyhart, 1995.

41. Breidbach, 2006.

schema, Haeckel offered many evolutionary trees aiming to represent actual evolutionary history.

But his view of evolution was not the same as Darwin's. It was a blend of concepts borrowed from Lamarck, Goethe, and *Naturphilosophie*. He didn't believe in the action of natural selection. His fundamental biogenetic law, according to which ontogenesis recapitulates phylogenesis—that is, an organism in the course of its embryonic development goes through the different states its adult ancestors went through during evolution—was derived from the models of Meckel and Serres referring to the existence of a natural scale of beings. Haeckel can be credited with attracting the attention of biologists to the relationship between embryonic development and evolution. However, he made his theory of recapitulation an absolute principle, asserting that all organisms go through an identical primitive stage of development—which he called "gastraea"—and that all develop in an identical way from embryonic layers, assertions that would later be shown to be false. The illustrations he provided to support these resemblances were done too quickly, with the same image being used to represent embryos of different species, which led to Haeckel being accused of falsification.[42]

Even more so than Huxley, Haeckel identified his battle for the theory of evolution with that for progress, against priests and tyrants. And so he came up against Virchow, who thought it was premature to teach the theory of evolution in schools. As he grew older, however, Haeckel moved away from science and focused increasingly on his philosophical concepts and his adherence to monism, a concept according to which there is only one substance in the world. For Haeckel, living matter encompassed the attributes of the spirit.

42. Hopwood, 2006.

Not all German biologists shared the same blind enthusiasm for the theory of evolution. As we have seen, Weismann extracted from the Darwinian theory the role of the use and nonuse of parts and the inheritance of acquired characteristics and, for that reason, assigned a major role to natural selection. Following Wilhelm Roux, he even had competition between the parts of the organism, and natural selection, play a role in embryonic development.

But, in contrast to Weismann, many German-speaking cytologists had doubts about the role of natural selection. The Swiss anatomist Kölliker believed in the existence of several different origins for living forms, and thought that there were leaps in evolution. A student of his, the Swiss zoologist Theodor Eimer (1843–1898), popularized the term "orthogenesis," introduced in 1893 by Wilhelm Haacke (1855–1912), to describe a directed evolution. Oscar Hertwig also rejected Darwinian theory. He was convinced of the existence of an inheritance of acquired characteristics, and didn't think that a theory based on chance variations was a scientific theory. Furthermore, he rejected the application of the Darwinian model to human society.

However, Darwinism received strong support from Anton Kerner von Marilaun (1831–1898), a professor of botany at the University of Vienna. He wrote a classic work on plant geography in the Danube region, showing that plant distribution depended on the climate and geology. He was interested in interactions between plants and insects, and described the chemical weapons used by plants to defend themselves against insects. He was convinced that these complex interactions couldn't be the result of a direct adaptation à la Lamarck, but on the contrary were the result of natural selection as described by Darwin.[43]

43. Hartmann, 2008.

The French scientific world did not receive Darwinism warmly.[44] Cuvier's fixism was still influential. Pierre Flourens, one of Cuvier's disciples, would prove to be a resolute adversary of Darwinism. Fixism was bolstered by observations made by the botanist Alexis Jordan (1814–1897) from Lyon. By studying in the minutest detail specimens of species that had already been described, Jordan showed their heterogeneity, and from that multiplied the number of existing species. Jordan's work thus supported the idea of the stability of species. It was received very favorably by amateur naturalists, to whom his example perhaps gave hope that they would discover new species!

And Darwin's reception in France was not made any easier by his first translator, Clémence Royer (1830–1902). A freethinker and feminist, she turned her long preface to the translation of 1862 into a screed against religion, a pamphlet in favor of allowing natural selection to play out among the human races, and a warning against the negative consequences of social services for weak individuals. She also added many notes in which she systematically eliminated the questions and doubts that Darwin himself had expressed. The French title—*De l'origine des espèces, ou des lois du progrès chez les êtres organisés* (On the origin of species, or on the laws of progress in organized beings)—was also different from Darwin's original. Curiously, in spite of all these changes, Darwin granted permission for her to publish a second edition, after asking her to make a few changes to the first. It took her publication of a third edition (without his permission) for author and translator to part ways and for Darwin to look for another publisher in France.

The acceptance of the idea of evolution in France was the work of a new generation of biologists, most of whom were graduates

44. Conry, 1974; Farley, 1974.

of the École normale supérieur (ENS), and who were led by Alfred Giard (1846–1908). Giard achieved a large body of work as a naturalist and entomologist, focusing in particular on phenomena of parasitism. In 1874 he founded the Wimereux marine station. He was named professor at the ENS in 1887, and in 1888 assigned to teach the first course in evolution at the Sorbonne. He presented his ideas on evolution in his book *Les controverses transformistes* (Transformist controversies) of 1904. Edmond Perrier (1844–1921), a student then professor at the ENS and chair of mollusks at the Muséum d'histoire naturelle, in 1888 wrote a book titled *Le transformisme*, in which he provided a very personal synthesis of the various transformist concepts.[45] Gaston Bonnier (1853–1922) was also a professor at the ENS. A botanist, he was the author of many works on flora. Félix Le Dantec (1869–1917), also a student at the ENS, worked at the Pasteur Institute. He is the author of several books of philosophy.

They all shared an acceptance of the theory of evolution, but felt that it was incomplete, because Darwin said nothing on the origin of variation. More loyal to Geoffroy Saint-Hilaire than to Lamarck, they thought that this origin was to be sought in the external environment. Marked by the experimental philosophy of Claude Bernard, they wished to create, alongside experimental physiology, an experimental transformism: to establish laws that would connect variations in the environment to those of organisms. Bonnier was the first to set off on this path, observing modifications that plants underwent when their environment was changed (temperature, altitude).

This movement, called neo-Lamarckism following the introduction of the term neo-Darwinism,[46] differed from other

45. Perrier, 1888.
46. Loison, 2010.

forms of Lamarckism, such as the one promoted by Cope, which retained from Lamarck the idea of an inner force and of the role of habits. The French neo-Lamarckians were materialists and sought the physicochemical mechanisms of evolution. The attenuated bacteria used by the Pasteur school to produce vaccines were, for them, the model of experimental transformism.

Changes induced by the environment are transmitted to descendants: neo-Lamarckism rests entirely on the inheritance of acquired characteristics. The experiments of experimental transformism would never give convincing results. And the neo-Lamarckians were never able to respond to Weismann's objections. French neo-Lamarckism would slowly fade away in the first half of the twentieth century.

One French biologist, however, deserves to be mentioned: the amateur Gaston de Saporta (1823–1895), who was quickly won over by evolutionist ideas. He corresponded with Darwin, and made many observations on plant fossils from which he proposed hypotheses on the evolution of plants.

In general, the idea of evolution was widely and quickly accepted. The role of natural selection was debated. Many saw it as a secondary phenomenon, eliminating unadapted organisms, and not the driver of evolution. The heart of the question was the origin of variation.[47] It was seen variously as the direct result of the action of the environment, or as that of an internal force, or as the result of "leaps." The idea that evolution could be the result of small, random variations sorted by natural selection was not the dominant theory at the beginning of the twentieth century. In that context, then, there was indeed an eclipsing of Darwinism.[48]

47. Hoquet, 2009.
48. Bowler, 1983.

From Biogeography to Ecology

It is difficult to distinguish in the work of zoologists and botanists of the nineteenth century what today we call ecology, even though the term was beginning to be used.[49]

Alexander von Humboldt (1769–1859), following the publication of his account of his travels in Latin America and owing to his fame, had considerable influence. Between 1799 and 1804, accompanied by the French surgeon and naturalist Aimé Bonpland (1773–1858), he traveled across Amazonia, Cuba, and Mexico. *Le Voyage aux régions équinoxiales du nouveau continent* (*Personal Narrative of Travels to the Equinoctial Regions of America during the years 1799–1804*) would be published in 30 volumes between 1807 and 1834.[50] Humboldt was both a man of the Enlightenment, studying the Earth's magnetic field and volcanoes as well as plants and animals, and belonged to the new type of specialized scientist of the nineteenth century. He is considered the father of geography for his descriptions of the relationships between human beings and their environment, and the comparisons he constantly made between localized observations and general models. He also contributed to progress in the geography of plants, to which volume 27 of his travel account is entirely devoted. During his attempt with Aimé Bonpland to climb Mount Chimborazo in the Andes, he observed changes in vegetation related to the gradients of temperature.

Augustin Pyrame de Candolle (1778–1841), a botanist from Geneva, attempted to create a list of all known plants, a work carried on by his son, Alphonse de Candolle (1806–1893). He

49. Deléage, 1991; Drouin, 1991; Acot, 1998.
50. Humboldt and Bonpland, 1895.

classified his list following a "natural" method different from that of Linnaeus, one inherited from Bernard and Antoine-Laurent de Jussieu. He also demonstrated the importance of light to flora. Observing the competition among plants to occupy the same land, he spoke of a "war of nature" and of a "war of plants," expressions that were found later in Darwin and Wallace. He contributed to separating resemblances through homology (which today we attribute to descendance) and those by analogy linked to lifestyle (which we would consider the result of an evolutive convergence). He also described the circadian movement of the leaves of certain plants, and advanced the hypothesis that this movement is controlled by a plant's internal clock.

The notion of "plant communities" assumed growing importance in the scientific literature. A botanist from Göttingen, August Grisebach (1814–1879), compared these associations to different physiognomic characteristics. And Alphonse de Candolle attempted a classification of these communities based on physiology.

It was in the context of debates on evolution that ecology found a second wind in the middle of the century. In his *Principles of Geology* (1830), Charles Lyell had already envisioned the consequences that the introduction of a new animal species (like bears) might have on the species living in a territory (in this case, Iceland). Recall also Wallace's important work that aimed to establish the relationship between biogeographic distribution and the evolutionary history of organisms.

The *Origin of Species* contains several pages in which Darwin considers the chain reactions that the disappearance of a species may have on other living species. Beyond those emblematic examples, the entirety of Darwin's work shows the complexity of the relationships that exist among living beings and between organisms and their environment: the role of insects in the fertilization of plants, or that of earthworms in the renewal of soil necessary for plant development, for example. One of Darwin's great merits was to have shown that the apparent stability of different animal and plant populations in nature is the result of a fragile dynamic balance between an elevated rate of reproduction and an equally elevated mortality rate.

In 1866, Ernst Haeckel, a fan of neologisms, introduced the term "ecology" in his treatise *General Morphology of Organisms* to describe the science that looks at the relationships organisms have amongst themselves and with the external world. Haeckel is also often considered the father of political ecology for his arguments in favor of protecting nature.

At the same time, and independently, the industrialization of agriculture and the increasing use of fertilizer were accompanied by quantitative studies on the needs of plants, and in particular their mechanisms of absorbing nitrogen, which were the work of the German chemist Liebig and the French chemist Jean-Baptiste Boussingault (1802–1887).

The concept of the ecosystem appeared under the term "biocenosis" (life in common) proposed by Karl Möbius (1825–1908) and under that of "microcosm" advanced by Stephen Alfred Forbes (1844–1930).

Möbius, professor of zoology at the University of Keil, was asked in 1869 by the authorities in the Schleswig-Holstein region to look into difficulties being encountered by oyster producers. At the culmination of a comparative study undertaken in France (in the Marennes region) and in Great Britain, Möbius reached the conclusion that the problem came from overexploitation caused by the arrival of the railroad, and of a growth of the market resulting from it. He showed the cascading consequences of such overexploitation, up to the colonization of the territory by other animal species. In 1877 he introduced the term "biocenosis" to describe the community of life that overexploitation destroyed.

In 1887, the American zoologist Stephen Alfred Forbes, during his study of the lakes of Illinois, introduced the term "microcosm" to describe the entirety of trophic relationships among living species in the lakes, and described the consequences that a perturbation could have on the entire system. The first exhaustive description of an ecological system—in its trophic dimension, with a description of food chains—as well as a physical system, was by François-Alphonse Forel (1841–1912), a professor at the University of Lausanne, for Lake Geneva. He is the creator of limnology, the equivalent for lakes of what oceanography is for oceans.

These studies joined those of microbiologists such as the Russian Sergei Winogradsky (1856–1953) who no longer investigated the pathogenic role of microorganisms, but rather showed their essential contributions to the production and recycling of organic matter in soil.[51] Winogradsky first worked in Saint Petersburg, then at the Pasteur Institute in Paris toward the end of his life. He was the first to demonstrate the existence of chemoautotrophic microorganisms—that is, those capable of drawing their energy from composite minerals such as hydrogen sulfide.

Acknowledging these developments, in 1893 the physiologist and microbiologist John Burdon-Sanderson (1828–1905), president of the British Association for the Advancement of Science, raised the science of ecology to a ranking within biology equal to that of morphology and physiology. As in the time of Linnaeus, what would become the ecosystem was conceived in a stable equilibrium, an equilibrium that could, however, be al-

51. Ackert, 2006, 2007.

tered by human action. Ecology had not yet integrated the temporal dimension that the theory of evolution was beginning to instill into the living world.

Historical Overview

A Moving History

The active locus of scientific work continued to shift throughout the nineteenth century. Whereas Paris and France assumed a major role in the first decades with Lamarck, Geoffroy Saint-Hilaire, and Cuvier, from 1830 Great Britain, with Darwin, and the German-speaking countries with cytology and the beginnings of a science of heredity picked up the baton. The influence of Paris at the beginning of the nineteenth century can be easily explained: the spoils of war from the Revolution and the Napoleonic campaigns enabled the Muséum d'histoire naturelle to amass a collection of material that until then had been spread throughout Europe. French science, and in particular biology, gained strength through military conquest. In contrast, later developments—the difficulty in accepting the theory of evolution and the rejection of natural selection, the limited place of cell biology, and the lack of any reflection on the mechanisms of heredity—deserve an explanation. The influence Cuvier still wielded through the work of his students is one of the components. The successes of Pasteur and of Claude Bernard are, paradoxically, another. Not only because they strongly supported French biologists' illusions of their superiority, but also because some scientific approaches were put forth as universal scientific methods—for example, the experimental method, of which Bernard and Pasteur were considered the champions—thereby devaluing scientific work that didn't conform to that model, such as the observations of cytologists and evolutionary scientists. Defining a scientific method, as Bernard explicitly did, can have counterproductive normative consequences.

The Birth of a Science of Heredity

This was one of the major events of the nineteenth century, even if it was the twentieth century that would assign it the name "genetics," by which we know it today. We have seen how it was born, simultaneously but independently, out of the work of Mendel on hybridization and from Darwin's investigation of the transmission of evolutive variation. Although Mendel had little to say on the mechanism, he was innovative in two ways: first, by showing the possibility of establishing laws of inheritance without knowing the embryologic mechanisms on which the formation of characteristics rests; second, by considering the organism as a set of characteristics, a level of organization not incompatible with seeing it as made up of the tissues and cells that biologists had recently learned to work with. Two problems remained throughout the twentieth century: how to connect the transmission of characteristics and their reproduction, and how to connect characteristics to other levels of organization of the living being.

The birth of a science of heredity might seem belated when hereditary phenomena had been observed since ancient times, and hereditary disease had been defined in the Middle Ages. The historians Staffan Müller-Wille and Hans-Jörg Rheinberger see it as the result of a set of practices that changed the relationship that living things (including humans) had with their environment.[52] Up until the eighteenth century, and including the

52. Müller-Wille and Rheinberger, 2007, 2012.

ideas of Buffon, it was possible to imagine that the characteristics of organisms were connected to the specific regions in which they lived. The acclimation of plants and animals that had come from newly explored lands, the displacement of human populations through colonization or slavery, showed that heredity was independent of the environment. The development of practices of artificial selection contributed to this biologization of heredity. The work of cytologists and evolutionary scientists also encouraged the idea that the construction of a science of heredity should be a priority. Similarly, at the beginning of the twentieth century, genetics would find support in the new practices of plant and seed producers, and in the care they henceforth took to keep track of the stages and crosses that had led to new varieties.[53]

Between 1810 and 1830, French doctors extended the notion of heredity, which had been limited since the Middle Ages to the transmission of a few diseases, to make it a central concept in an explanation of the moral and physical makeup of individuals.[54]

Biology: A Comparative Science, according to Auguste Comte

When, in his course on positive philosophy, August Comte made the comparative approach specific to the biological sciences, he was faithfully reflecting the development of those sciences: anatomy had become comparative, as would embryology and physiology a little later. There were multiple objectives for these comparisons. At the end of the eighteenth century, the

53. Thurtle, 2007.

54. Lopez-Beltran, 2004.

anatomist Félix Vicq-d'Azyr had compared the anatomy of different organisms in an attempt to discover what in the structure of an organ was important for its function. The comparison of different auditory organs revealed which structures were indispensable for hearing, and which were either accessory or linked to a particular characteristic of the auditory function. When Merkel compared the embryonic development of organisms, the comparison was in relation to those organisms' position on the chain of being. The same comparative method enabled Haeckel to place organisms on the evolutionary tree. Similarly, today, comparison of sequences of nucleic acids or of proteins can enable us to determine the genetic distance between organisms; in a protein it also enables us to define the parts that are essential to its function (corresponding to the most highly conserved sequences). The use, and interpretation of the results, of the tool of comparison can therefore have many ends. For twenty-first-century biologists, as for those of the nineteenth century following Haeckel, comparison is almost automatically placed in an evolutionary framework—to the point of sometimes forgetting that it continues to have other uses. It is within this context that a distinction between homology—resemblance linked to descendance—and analogy—resemblance linked to commonality of function—has had trouble emerging. Even today, an imperfect distinction between these two concepts fuels many debates.

Darwinism and Ecology: A Complex Relationship

Many people seem to connect the two disciplines. The observations of biogeographers prepare the way for an evolutionary interpretation of living phenomena. Darwin's observations on the fertilization of plants by insects, and more generally on the complex relationships among different species, fall under ecology, as much if not more so than under his description of evolutionary mechanisms. And the term "ecology" was introduced by Haeckel, a confirmed evolutionist.

And yet, similar descriptions may have different objectives. For ecologists, beyond the providentialism of Linnaeus, it was a matter of explaining how the different interests of organisms contribute, if not to harmony, at least to an equilibrium. The choice of the systems studied shows that it is their stability that must be explained, the *explanandum*. For an evolutionist, these complex interactions are the driver of evolution, that which orients the continually produced selection of random variations. We should add that the pace of the evolution of organisms and of that the transformation of ecosystems are very different.[55]

Ecology and evolutionary biology share not only an extended period of birth, but also a contradiction. They are two realms of science that aim to describe nature as it is, or even as it was before the perturbations introduced by human activity. However, the decisive stages in the formation of these sciences, the emergence of the notion of the ecosystem and of Darwinian theory, are the fruit of human action, and of the study of the perturbations that it causes. In the first case, the overexploitation of oyster beds and overfishing engendered the notion of a community of life. Most of the observations Darwin assembled in support of the theory of natural selection originate in human activities: the work of breeders, the effects of the introduction of plants and animals in regions where they were previously absent, and so on. This is an opportunity to remind ourselves that scientific knowledge is most often the fruit of voluntary human action, rather than a passive observation of the world.

55. Delord, 2009.

Biogeography

Biogeography is the science that studies the distribution of plants and animals on the Earth's surface. It is one of the roots of ecology, and some biogeographic observations found an explanation with the advent of the theory of evolution.[56] It has provided observations that have favored the development of the idea of "living communities," demonstrated the influence of the physical environment on the nature of animal and plant populations, and thus helped the emergence of a physical ecology. It has also revealed distributions of species that are not explicable by just the effects of climate, and thus opened the way to a history of the populating of Earth.

Buffon, in 1761, was the first to notice that mammal species are different in the Old and New Worlds, and deduced from that what many have called "Buffon's law," according to which the distribution of species is explained by the existence of natural geographic barriers. Humboldt and Augustin de Candolle would extend that law to all animals and plants.

These observations preceded and paved the way for the development of the theory of evolution—in the same way that the observations of embryologists anticipated the existence of evolutionary relationships among organisms. They would be reinterpreted by Darwin and Wallace as the result of two processes, evolution through natural selection and migration (which these authors called "dispersion"). Migration could be "improbable," enabling, for example, a species to cross the space separating a continent from an island.

In the twentieth century, biogeography would face another challenge: incorporating the results of plate tectonics into its

56. Nelson, 1978.

explanations. This additional cause for the geographic distribution of species—the drifting of land masses—was added to the two previously described causes, evolution through natural selection and migration. Evolutionists, such as George Simpson, rejected the idea of drifting of continents. In contrast, the biogeographer Léon Croizat (1894–1982) tried in vain to make the shifting of land the main if not single cause for the distribution of species on the surface of the globe.

Biogeography is an excellent example of the recurring difficulties biologists experience in connecting and/or prioritizing the multiple causalities at the origin of living phenomena.

The Epistemological Originality of the Darwinian Model of Natural Selection

Darwinian theory, like Lamarckian or neo-Lamarckian theories, aims (among other things) to explain the adaptation of organisms to their environments. But while adaptation is a more or less direct consequence of the action of the environment in Lamarckian theories, it is only one indirect consequence in Darwinian theory: adaptation is only the result of the greatest proliferation of the best adapted organisms.

The originality of the Darwinian explanation means that it is impossible to classify it among the types of explanations most commonly used in science. Darwinism doesn't rely on laws, which explains why so many physicists didn't embrace it (in the nineteenth century, but even today), so accustomed are they to explaining natural phenomena by the existence of laws. Nor is Darwinism a mechanist explanation: there is no comparison with any machine. Nor is it mechanist in the sense that philosophers of science have recently attributed to the term: an explanation calling upon entities and activities exerted on those entities.

This epistemological strangeness is often not stressed enough, because it is the source of much of the rejection and misunderstanding of Darwinian theory that are falsely attributed to metaphysical or religious opposition.

The prolonged opposition of many paleontologists to Darwinian theory also has epistemological roots.[57] The fossil record, by its very nature, favors a nongradualist view of evolution. One of the possibilities offered to paleontologists is to compare fossil forms that succeed each other over time. This type of comparison, different from those described above, leads to a search for laws of transformation of those forms, such as the laws of Cope seen briefly above. From the discovery of these laws to the search for a simple mechanism, internal to the organism, there is but one step, which is rapidly taken. It is not impossible to derive these laws from a continuous action of natural selection, for example, but the path is longer and seems less obvious.

Science and Religion

In the difficult relationship between science and religion, science and faith, Darwinian theory occupies a prominent place, alongside the trial of Galileo—but a much more important place in current debates than the latter.

The confrontation that began in the middle of the nineteenth century went far beyond the question of the mechanisms of evolution. The battle in favor of Darwinism was—for Huxley, Haeckel, and many others—a battle for progress, and against political and religious conservatism. To those two general aspects can be added two elements that are specifically tied to the

57. For this opposition, see Wolf-Ernst (1986).

nature of theological discourse at the time: a still literal reading of the Bible by many Christians, and the importance in England of the natural theology of William Paley (1743–1805). For Paley, an observation of nature directly revealed the work of God—just as the presence of a watch demonstrated the existence of a watchmaker, the existence of living beings was the mark of the action of a creator.

This general context having been established, we must avoid combining these elements: not all Christians were opposed to Darwinian theory, and not all those who rejected it did so for religious reasons. Many religious biologists, like Asa Gray, accepted the Darwinian theory of evolution. And most biologists who rejected the theory did so for scientific reasons, convinced that natural selection was not a sufficient mechanism to explain evolution, or refusing the gradualism of the Darwinian model and opting for the existence of evolutive leaps.

There are a lot of similarities among the arguments exchanged at the time between the Darwinists and their religious opponents, and those advanced in contemporary debates. Those who oppose Darwin's theory for religious reasons often believe either that the Darwinian model doesn't explain evolutionary trends, or that it isn't able to account for the perfection of organisms. It is this second argument, of "design," that one finds in the words of contemporary proponents of "intelligent design."

Conversely, there is an element that played a minor role in the debates of the nineteenth century, and which gradually took on a major role in the twentieth century: the role of chance. It wasn't religious believers but other scientists such as Oscar Hertwig or the neo-Lamarckians who thought that calling upon chance was unacceptable. Although Darwin didn't ascribe a major role to chance, and even suggested in some passages of the *Origin of Species* that talking about chance is only a reflection of our ignorance.

Darwin and the Human Being

The preface written by Clémence Royer shows that, very quickly, ideas of competition and natural selection were being applied to the human species. Darwin makes only a few allusions to humans in the *Origin of Species*, but those allusions are sufficiently explicit that it can escape no attentive reader that the mechanism of evolution through natural selection also applies to humans. There were, nonetheless, two different readings of Darwin's book. The first, that of Herbert Spencer (1820–1903), was social, and retained from Darwin's message only the idea of competition. Interindividual competition is good for the progress of societies, for their ability to transform and innovate. The second was biological, and goes as follows: There are differences among human individuals, and up to now those have been sorted through natural selection and the disadvantageous variations eliminated, which enabled the progress of the human species. But natural selection no longer functions in our societies: the weak are taken in and cared for, and primitive populations civilized. The proof of this interruption in natural selection, as Galton shows, is that the number of children born no longer depends on either the quality of the parents or that of populations: in England, it is "superior" individuals who reproduce the least; and the English have many fewer children than do African populations. The eugenics promoted by Galton was a response to those changes, an appeal to the implementation of a form of artificial selection to replace natural selection. In its positive form, it was a matter of encouraging the reproduction of the "best"; in its negative form, it would prevent the "less fit" from reproducing. Eugenics primarily involved individual differences within a population. As for the differences between human populations, the preceding considerations inspired by Darwin's theory reinforced the racist ideologies that predated them. The theory of polygenism—that is, of the existence of several species of humans with different origins—was, as we have seen, supported by several eminent nineteenth-century biologists. It is impossible to know if it was racist prejudices that oriented scientific work, or theories and scientific observations that favored the development of racist theories. Since the Church was often the institution that took care of the weak, the battle in favor of Darwinian theory and for eugenics joined the one against the Church.

Two concepts, which we might through analogy call preformationist and epigenetic, have sought to account for this objective collusion between Darwinian theory, racist theories, and eugenics. The first holds that those consequences (racism and eugenics) are contained in Darwin's theory; the second, that it is circumstances, the ideological environment, that enabled some to use Darwinian theory for racist and eugenic ends.

Many historians have attempted to exonerate Darwinism by analyzing Darwin's own ideas. Darwin was opposed to slavery—he even had a heated argument with FitzRoy, the captain of the *Beagle*, on the subject. In his work *The Descent of Man, and Selection in Relation to Sex* he insists on a dramatic transformation that occurred in the human species with the emergence of sympathy, what today we would call compassion. Unfortunately, Darwin's attitude was more ambiguous than that described by the gilded legend that surrounds him. As we have seen, he made only belated objections to the nonetheless explicit preface by Clémence Royer. He described the disappearance of "primitive peoples," like the Maori, as the result of a natural phenomenon of competition, and compares that to the replacement of indigenous rats by European rats. He accepted the necessity of a eugenic policy such as the one proposed by Galton, although he hoped that individuals would voluntarily adapt the size of their families to their "quality."

Darwin devoted his life to demonstrating the reality of the biological evolution of the human being, and the importance of the mechanism of natural selection. Even though he saw the importance of sympathy—which for him was one of the highest values—in the human species, he could not renounce the supreme role of natural selection in human evolution. To do so would have been to join the camp of those, including Wallace, who considered the human being and its evolution to be "exceptions." The ideological and political battles into which the theory of evolution was plunged tolerated no nuance.

Contemporary Relevance

Epigenetics and the Return of Lamarckism

Epigenetic marks are chemical modifications of DNA and the proteins that surround it that cause a modification of gene activity. They can be induced by the environment and are sufficiently stable to be preserved during cell division and, in some cases, transmitted by organisms to their descendants.[58] Their involvement in "reprogramming" the nuclei that accompany cloning, as well as their possible role in some pathological processes such as cancer, explain the great amount of work currently being devoted to them. But the popularity of epigenetics goes beyond the circle of specialists: it is "in fashion." Its results

58. Jablonka and Lamb, 2005.

apparently show that genes are sensitive to the action of the environment, and thus also to human action, and that these modifications of the activity of genes can be transmitted to one's descendants—which would mean the end of genetic determinism, and a return to Lamarckism.

But which Lamarckism are we talking about? As we have seen, Lamarckism has several faces. What is proposed is neither a physicochemical mechanism of complexification as Lamarck understood it, nor some kind of spiritual force, as some American neo-Lamarckians have proposed. The concept that emerges from some current work in epigenetics might be closer to that of the French neo-Lamarckians—a direct effect of the environment on organisms—or might be related to Lamarck's hypothesis that "habits" can be transmitted to one's descendants.

However, to associate them with Lamarckism or with neo-Lamarckism, these modifications would have to enable organisms to adapt to their environment. When addition of a chemical compound modifies the epigenetic marks not only in an individual organism but also in its descendants, as has been observed in a number of experiments, nothing indicates that those modifications are adaptive.

Attributing an evolutionary role to these epigenetic marks also comes up against the same type of objection Weismann had to the inheritance of acquired characteristics: not because of the problem raised by the separation of the soma and the germen and the isolation of the germinal lineage, but because, despite repeated assertion of the existence of an inheritance of acquired characteristics, very little experimental evidence can be offered in its favor. Cases of transmission of epigenetic marks from one generation to another are rare in the scientific literature, and in most cases could be the result of a reproduction of those marks in the following generation, and not of a direct transmission.

Contrary to what Weismann thought, the separation between the soma and the germen does not exist in all organisms—in particular, not in plants. It is perhaps not by chance that epigenetic marks seem particularly abundant in them, and that cases of intergenerational transmission of those marks are most easily demonstrated in them.

Compensation and Life Histories

One idea that many nineteenth-century biologists shared—Darwin as well as Geoffroy Saint-Hilaire—was that of compensation. If, during evolutionary transformations, an organ becomes more developed, that growth would often be compensated by a diminution of another organ. This view was obviously connected to Kant's concept of the organism, as well as to Cuvier's law of the correlation of parts. The explanation given for this notion is that there is a general balance sheet within an organism: a growth must be compensated for by a reduction.

This idea was picked up again in the twentieth century by evolutionary scientists. Two species that are close on the evolutionary tree may have very different "life histories:" one might have a long life but a low fertility rate, whereas the other has a short life, but a high fertility rate. The explanation given is that organisms possess a fixed amount of "energy" that they can invest either in the prolongation of life or in reproduction. The two objectives are not independent—one must live to reproduce!—and, according to their specific environment, organisms adopt the life history that is optimal to ensure the best rate of reproduction, to increase their "fitness," as evolutionary scientists say. A more precise version of the notion has been proposed by Thomas Kirkwood (1951–) to explain the investment necessary for a prolongation of life: organisms are continually damaged by their environment or their own functioning. Repairing this damage has a price, taken from the same budget as that which serves for reproduction.

Another theory of aging, proposed by the geneticist George Williams, also implies an antagonism between reproduction and lifespan. Mutations that increase the efficiency of reproduction are pleiotropic—that is, they have multiple effects, in particular that of inducing dysfunctions that shorten life at an advanced age when the reproductive period is largely over. In terms of "fitness," the benefits they provide are superior to their costs, which are limited to the age in which they produce their harmful effects.

What is rather curious is that neither twentieth-century biologists nor those of the nineteenth century felt the need to justify the necessity of this compensation, or did so only by using economic or energy-related metaphors. Why, for example, are mutations that increase the fertility rate necessarily pleiotropic? Might we not imagine that they have only one (beneficial) effect, or even beneficial effects on both fertility rate and lifespan? The origin of this view can clearly be found in the concept of an economy in nature, such as Linnaeus described, but which for him was explained by the benevolent action of the Creator. Let's mention, however, that the first evolutionist theory of aging, proposed in 1952 by Peter Medawar, simply explained the phenomenon as the weakening of natural selection beyond reproductive age.[59]

The End of Orthogenesis?

The notion of the existence of orthogenesis—the apparently directed or predetermined evolution of an organ or a structure—comes primarily out of paleontological observations, where if often appears as though a structure, over a long period of

59. Medawar, 1952.

evolutionary time, has undergone transformations all going in the same direction.

Orthogenesis has been considered an illusion connected to the incomplete nature of the fossil record. But it has also been considered real, explained by the action of natural selection in a constant environment. And it has been justified by the existence of constraints on embryonic development that would channel evolutionary development. Finally, it has been interpreted, as among the American neo-Lamarckians or in the writings of Teilhard de Chardin, more "metaphysically," relating to a "directed" evolution.

Contemporary biologists do not consider orthogenetic phenomena to be major evolutive facts, even if, as we have seen, they might be explained in a way that is completely compatible with current evolutionary theory. They are reminiscent of the theories that, from the nineteenth century until the middle of the twentieth, gave meaning to history (of human societies). In biology, they are contrary to an open evolutionary history as Gould sketched it in the twentieth century.

Which isn't to say that the idea of an oriented history has completely disappeared. It has taken on other forms, and is expressed with other words that mask this continuity. For example, no one objects when it is asserted that a country has advanced in the reform of its institutions, its legislature, or its economy, whereas such a declaration suggests that such transformations go in a well-defined direction, which is rarely the case.

Did Geoffroy Saint-Hilaire Win the Argument with Cuvier?

Just before the turn of the twenty-first century, several articles that appeared in scientific journals caused a big stir. By describing the expression of genes involved in early embryonic development, both in insects and vertebrates (*Sog/chordin* and *dpp/bmp* genes), researchers showed that the dorsoventral inversion between insects and vertebrates—from which Geoffroy Saint-Hilaire had developed his hypothesis to justify the existence of a single plan of organization in animals—could be read directly in profiles of expression of these genes.

Among the authors and other commentators, there was an obvious pleasure in revealing that the great Cuvier was wrong, whereas Geoffroy Saint-Hilaire had told the truth. There ensued a complex debate on the precise mechanisms of development in these two types of organisms, which we will not go into here. The question is whether the dorsoventral turning is the main event, or if it is rather the consequence of a series of events—for example, cellular migrations—that occurred during development. The debate is still going on; specialists have been unable to offer a resolution.

There is a more fundamental issue. Is it possible to compare contemporary questions with those of 1830? Cuvier and Geoffroy Saint-Hilaire clashed on the existence (or not) of a single plan of organization—that is, that the nature of the parts of organisms and the relationships that connect them are identical in all animals. The current debate is based on the identity of genes and of their realms of expression. What connection can be made between genes and parts of an organism? Can we reduce the latter to the former? Can the homology of genes be extended to that of the structures in which they are expressed? Genes are often considered to be a simple toolbox that evolution has drawn from. It isn't obvious that the notion of "deep homology"[60]—resting on the existence of networks of developmental genes appearing very early during evolution—better

60. Shubin et al., 2009.

resolves the debate that opposed Geoffroy Saint-Hilaire and Cuvier.

The Mathematical Laws of Morphogenesis: The First Steps of Phyllotaxy

Phyllotaxy is the branch of botany that studies the arrangement of leaves on their stem.[61] It is the oldest domain of biomathematics and, for that reason, its history is interesting, although it has not often been studied.

Theophrastus and Pliny the Elder, like the Renaissance authors Andrea Cesalpino and Leonardo da Vinci, had already noted the existence of regularities in leaf arrangement—leaves growing in opposing pairs, or forming a spiral. In the eighteenth century, Charles Bonnet provided the first precise description of these spirals.

In the 1830s, a group of scientists including Karl Schimper (1803–1867) and Alexander Braun (1805–1877) completed Bonnet's description by revealing that the spirals could be described by a Fibonacci sequence. Around the 1870s, Wilhelm Hofmeister and Julius von Wiesner (1838–1916) proposed two different mechanisms to account for their formation.

Hofmeister suggested that the preliminary leaves (the primordia) appear regularly at the top, where the space left by preceding primordia is greatest. In 1878, Simon Schwendener (1829–1919) proposed a slightly different theory, suggesting that the arrangement of leaves was the result of the contact pressure that each primordium exercised on its neighbors.

61. Montgomery, 1970; Adler et al., 1997.

In 1875, Wiesner advanced a radically different explanation, calling on Darwinian theory: the spiral arrangement of leaves is optimal for their exposure to light. He carried out experiments to attempt to confirm that hypothesis.

In 1882, the great plant physiologist Julius Sachs denied the value of any mathematical model to account for this regularity in the arrangement of leaves on their stems. His influence closed the debate for almost a century, until the 1970s, when a great deal of work was published on the mathematical modeling of phyllotaxy.

The cultural context in which this work was carried out is important. In 1830, *Naturphilosophie* was still dominant in German universities, and a search for a mathematical expression of the forms of organisms naturally found its place there. By contrast, in the 1870s, plant physiology was developing both in reaction against Linnaean botany and against the legacy of *Naturphilosophie*.

By the beginning of the 1880s, the debate had solidified: two different explanations—one mechanist, the other Darwinian—were proposed to explain this regular arrangement of leaves, and the value of a biomathematical approach was severely contested. As we have seen, work in this area has resumed in the past few decades, but the dividing lines that appeared at the end of the nineteenth century remain.

Another Mendel?

We have mentioned the different legends that surround Mendel's life and work. Some might be explained by the dramatization that always surrounds the lives of great people. Others stem from ignorance: about the training Mendel received, the nature of the religious order to which he belonged, or about the position of Brno, which was close to one of the greatest European capitals.

But there is something else: Mendel's behavior didn't correspond to what we, observers from the twenty-first century, might expect from a scientist. Let's leave aside Mendel's religious convictions, which were much used by communists in the twentieth century to discredit genetics. The way Mendel turned away from his work on plants and devoted the greater part of his time to managing the monastery is shocking—scientific research demands that it be granted first place, if not the only place, in the lives of those who engage in it.

Mendel is not a unique case among scientists. Newton devoted a lot of time to alchemy, Cuvier to politics, and Haller to diplomacy. Rather than looking for fallacious reasons for Mendel's behavior—such as an unfavorable reception of his work, which had supposedly discouraged him—perhaps it would be more useful to question the caricatural image we have of scientists. It is a false image, often promoted by scientists who have abandoned daily contact with research for administrative and managerial responsibilities. It is a restrictive image, because it causes regret among those who, for one reason or another, have had to abandon research—regret that may be difficult to bear. Finally, it is a socially dangerous image, because it surreptitiously introduces a hierarchy of human activities.

8

The Twentieth Century (Part I)

THE DIVERSITY OF FUNCTIONAL BIOLOGY AND THE BIRTH OF MOLECULAR BIOLOGY

The Facts

The first 60 years of the twentieth century saw a great diversification of biological disciplines—the advent of biochemistry, genetics, virology, ethology, and others—and two unifying developments: the emergence of molecular biology and that of modern evolutionary synthesis. We can return to the distinction we made for the nineteenth century between functional biology—biochemistry, physiology (this chapter)—and a more holistic biology, focusing on the study of the relationships among living beings, and of living beings with their environment (chapter 9), although some disciplines, such as genetics and ecology, can move between the two. The final chapter will look at what happened after the 1970s, in particular at the consequences of the encounter—sometimes the confrontation—between the molecular view of biology and the principal biological disciplines.

Biochemistry

BÜCHNER'S EXPERIMENT AND DECIPHERING METABOLIC PATHWAYS

Every description of the history of biochemistry agrees that the German physiologist Carl Neuberg (1877–1956) invented the term in 1903, and that the cell-free alcoholic fermentation of sugar experiment carried out in 1897 by the German physiologist Eduard Büchner (1860–1917) was the discipline's scientific foundation.[1]

History isn't quite so simple. The term "biochemistry" had been used on multiple occasions since the 1880s. And Büchner's experiment was fundamental only if one attributes to the presence of living organisms in fermentation the importance that Pasteur acknowledged. Furthermore, many researchers had attempted to perform Büchner's experiment, such as the Russian physician Maria Manaseina (1841–1903) in 1872, but the results they obtained were not convincing.

Above all, the experimental approaches of biochemists during the first years of the discipline's existence were little different from those that physiologists had used before them.

Büchner's experiment is nonetheless important in that it refined new techniques for breaking up yeast cells without damaging their contents, and above all for the possibility that it opened up—which would be effectively achieved in the following years—to break down, through working on cell extracts, the chemical stages of complex biological processes such as fermentation. Certain experimental systems were favored for this work, such as yeast extracts and, especially, muscle extracts, as

1. The information in the "Biochemistry" section is drawn from Kohler (1975, 2008) and Fruton (1999).

muscular activity is directly associated with the complex processes of fermentation and respiration.

Three names are emblematic of the successes of this branch of biochemistry during the first decades of the twentieth century. The first is Otto Meyerhof (1884–1951), who differentiated the two functional phases of so-called "rapid muscles": the production of lactic acid during muscular effort through the breakdown of stored glycogen, followed by a partial oxidation of the lactic acid during the recovery phase. Gustav Embden (1874–1933) gradually completed a description of the steps in the conversion of glycogen to lactic acid—a process called glycolysis, or the Embden-Meyerhof pathway. One of Meyerhof's students, Karl Lohmann (1898–1978), at the same time—but independently of Cyrus Fiske (1890–1978) and Yellapragada Subbarow (1896–1948)—characterized ATP, the energy currency of the cell, the molecule that provides energy both for the synthesis of complex molecules and for muscular contraction and the production of neural impulses.[2]

The second name is that of Otto Warburg (1883–1970), a student of Emil Fischer, who also worked in Berlin, then in Munich, and who characterized and purified the enzymes involved in the processes of cellular respiration. Warburg developed or implemented the equipment necessary for his work: the "Warburg apparatus" allowed precise measurement of the gases consumed and released during the reactions of respiration, and the spectrophotometer (used before him by David Keilin [1887–1963]) of the absorption of light by the molecules involved. He also studied the mechanisms of photosynthesis. In 1924, he discovered that cancer cells ferment in the presence of oxygen, which neither muscle nor yeast cells do. He saw this alteration

2. Maruyama, 1991.

of metabolism as the cause of cancer, and adamantly defended that hypothesis for the rest of his life.

The third name we will mention is that of Hans Krebs (1900–1981), a student of Warburg. Krebs showed that certain metabolic transformations corresponded to cyclical sets of reactions—including the formation of urea (1932), and the reactions intermediate between the fermentation pathway deciphered by Embden and Meyerhof and the respiratory processes studied by Warburg—now known as the Krebs cycle (1937).[3]

These discoveries were small links in the very long chain of work that would lead to the deciphering of all metabolic reactions. A satisfactory explanation of the processes of oxidation wouldn't be proposed until the 1960s, and wasn't accepted until the end of the 1970s. This work, which aimed to characterize the molecules and the transformations of energy involved (and, in particular, the accompanying release of heat) is reminiscent of that of Lavoisier, but differed from the latter's work by the use of much more precise techniques for measurement and analysis.

Two of these three researchers, Krebs and Meyerhof, being Jewish, had to leave Germany in 1933 (Krebs) and in 1938 (Meyerhof); Krebs went to Sheffield in England and Meyerhof to Paris and Philadelphia. The history of biochemistry bears the traces of the turbulence of the twentieth century.

THE STUDY OF MACROMOLECULES

Büchner's experiment tore down the divide that had existed until then between two types of biological catalysts: enzymes

3. Holmes, 1991 and 1993.

(which are soluble) and yeasts (which are living organisms).[4] The catalytic activity of yeasts turned out simply to be due to the enzymes they contain.

The characterization of the precise chemical nature of enzymes—whose catalytic activity is essential to the performance of the various steps of metabolic pathways, which at the time were being deciphered—became a primary goal. The predominately protein nature of enzymes had been known for a long time, but the precise nature of the catalyst was still a mystery. Did it involve, as the German chemist Richard Willstätter (1872–1942) and the Pasteurian Gabriel Bertrand (1867–1962) believed, small molecules associated with those proteins, or was catalytic activity carried out directly by proteins? This debate overlaid another that was taking place at the same time, one that pitted those, like German chemist Hermann Staudinger (1881–1965), who thought that proteins are macromolecules—that is, large molecules—against the many physicochemists who were convinced that the interior of living cells comprised a complex assemblage of small molecules organized in the form of colloids.

The answer would be provided experimentally. In 1926 at Cornell University, James Sumner (1887–1955) crystallized the enzyme urease (so called because it was capable of breaking down urea), extracted from beans. He showed that the crystal contained only a pure protein. He repeated the experiment again in 1937, crystallizing catalase. At the time, obtaining crystals was a great achievement, considering the techniques available, all of which came from organic chemistry. In 1929, John Northrop (1891–1987), at the former Rockefeller Institute for Medical Research in New York, purified and crystallized pepsin, followed

4. The information in this subsection is drawn from Tanford and Reynolds (2001).

by trypsin, chymotrypsin, and their precursors. Each time, the enzyme was revealed to be a protein. Unlike James Sumner, Northrop used new techniques to purify the enzymes—ultracentrifugation and electrophoresis—whose development accompanied the rise of molecular biology, as we will see below.

The irony of history is that for its catalytic activity, urease needs a metal—nickel—which Sumner wasn't able to observe, because the metal is present only in trace amounts, undetectable using the techniques of the time, in the solutions used to assay the activity of urease.

The description of other macromolecules of life—carbohydrates, lipids, nucleic acids—also progressed slowly during the same period. The bonds between the base, sugar, and phosphate in nucleotides were identified. The distinction was made between RNA and DNA (through the nature of the sugar present), and the proportion of the different bases in DNA measured. These results led to as many erroneous simplifications as they did to precise data. It was proposed that DNA was specific to the animal world, and RNA to the plant world; another theory saw DNA as a monotonous repetition of the four bases—the "tetranucleotide" hypothesis, proposed by the American chemist Phoebus Levene (1869–1940).

Analytical techniques progressed with the development of specific reagents that would react with the various types of macromolecules—Folin's reagent for proteins, and Feulgen stain (1924) for nucleic acids. Enzymes, partially or completely purified, became tools for measuring or characterizing other components of living things: before its crystallization by Sumner, urease was already being used to assay urea.

The different stages of enzymatic catalysis were described by the French biochemist Victor Henri (1872–1940) in his 1903 work *Lois générales de l'action des diastases* (General laws of the action of diastases), and later in 1913 by Leonor Michaelis (1875–1949) and Maud Menten (1879–1960).[5] These descriptions paved the way for the development of assays for enzyme activity, and for representations of that activity that would enable its quantitative study. Even more important was the work of physicochemists on the acidity of solutions (pH), ionic strength, and the effects of temperature, which enabled the preparation of solutions that were optimized for the purification and measurement of enzyme activity, and thereby a reduction of what had, since the middle of the nineteenth century, been a hindrance to that work: a spontaneous loss of activity.

VITAMINS

Another realm of research within biochemistry that we have not yet discussed involves the study of vitamins. Although this work was comparable to the earlier research in physiology, it nevertheless helped to enhance the visibility of biochemistry and to convince a large audience of its usefulness.

The discovery of vitamins came out of a complex history that spans several centuries.[6] In antiquity, the Egyptians recommended eating liver to fight against certain diseases of the eyes (due, we now know, to a vitamin A deficiency). During the winter of 1535–1536, three ships belonging to Jacques Cartier (1491–1557) became stuck in the ice of the Saint Lawrence river. A lethargy accompanied by a swelling of the limbs afflicted a large number of the crew, who were then miraculously cured by a concoction prepared by Indians out of the stalks and leaves of a local plant. In 1747, the Scottish doctor James Lind (1716–1794) noted

5. Deichmann et al., 2013.
6. Ihde and Becker, 1971.

that the illness—scurvy—which struck ship crews deprived for several weeks of fresh food, could be prevented by eating lemons.

In the 1880s, the notion of deficiency diseases—those resulting from the lack of an essential substance—gradually emerged. In 1881, Nicolas Lunin (1853–1937), working in Basel, observed that whereas milk was a complete food, a diet consisting of the known components of milk did not allow mice to survive. The first work on beriberi, the deficiency of vitamin B1, was done by Kanehiro Takaki (1849–1920) in 1884. Umetaro Suzuki (1874–1943) defined the notion of vitamin in an article published in Japanese in 1910, but its bad translation prevented its originality from being appreciated.

Official history, and in part the Nobel committee, has retained three names in the discovery of vitamins. The first is that of the Dutch physician Christiaan Eijkman (1858–1930).[7] Stationed in Indonesia, he noticed that chickens fed on unpolished rice (still with the seed hulls) had no sign of illness, whereas chickens fed on shelled white rice presented symptoms analogous to those of beriberi. His first idea was that beriberi was due to a toxin, and that the shell of rice contained an antitoxin. Several researchers claimed, moreover, to have isolated the microbe of beriberi. It would take Eijkman some time to abandon Koch's and Pasteur's model of the illness, and to realize that it wasn't caused by an added element—infection from a microbe—but from a deficiency, the lack of a substance, provided by food and essential to life.

The term "vitamin" was introduced in 1912 by the Polish biochemist Casimir Funk (1884–1967) to describe the active element that he believed he had extracted from rice hulls and which prevented the onset of beriberi. The term means a substance essential to life that, from a chemical point of view, is an amine (thiamin, later named vitamin B1).

7. Carpenter, 2000.

The same year, the English chemist Frederick Gowland Hopkins (1861–1947) defined vitamins more generally as elements that are essential (in very small quantities) in the diet, based on research he carried out in the preceding years—research in every way comparable to that conducted 20 years earlier by Nicolas Lunin.

In the following years, Hopkins and others isolated other vitamins, characterized them, and demonstrated and sometimes explained their role (in particular as coenzymes, small molecules essential to the action of enzymes). This attribution of a specific role—catalytic—to vitamins would contribute to the recognition of a new category of active substances in organisms, which until then had not had a place in the explicative framework established by physiologists.[8] Vitamin A was purified in 1917, and vitamin D ("antirachitic," which prevents rickets) in 1919. In 1931, a chemist of Hungarian origin, Albert Szent-Györgyi (1893–1986), established the chemical structure of vitamin C and confirmed its preventative action against scurvy. In 1933 in Otto Warburg's lab, George Wald (1906–1997) demonstrated that a derivative of vitamin A is present in the retina, explaining the nature of disorders resulting from a deficiency of that molecule.

Endocrinology and Neurophysiology

THE DESCRIPTION AND CHARACTERIZATION OF HORMONES

Anatomists and physiologists have known about glands for centuries. Although the function of glands with an excretory duct (exocrine glands) had been gradually deciphered, that of glands without such ducts for a long time remained unknown. The

8. Braun, 2011.

discovery of their method of secretion, via the blood ("endocrine"), and the characterization of their products (hormones) became a very lively field of research at the beginning of the twentieth century.

The history of hormones has a lot in common with that of vitamins. Observations went back to ancient times, but the rapid advances in work on hormones at the beginning of the twentieth century were the result of the progress made in organic chemistry. Research on hormones, like that on vitamins, benefited from great visibility, and the characterization of hormones led to hopes for rapid therapeutic uses, and solutions to certain social problems, such as those resulting from a "badly controlled" sexuality.

The characterization of hormones is closely connected to that of vitamins—cholesterol is the precursor of both vitamin D and sex hormones—and that of neurotransmitters, which we will discuss below: adrenaline is both a hormone and a neurotransmitter.

In 1849, Arnold Berthold (1803–1861), a disciple of Blumenbach at the University of Göttingen, had observed that castrating young roosters caused secondary sexual characteristics to disappear, but that those could be recovered when testicles were replaced in the animal. He proposed that the observed effects were due to an indirect action of the testicles on the blood.

After injecting himself with fluids extracted from the testicles of dogs and guinea pigs, the physiologist Charles Brown-Séquard (1817–1894), the successor of Claude Bernard at the Collège de France, in 1889 attributed to those fluids an effect of rejuvenation and a return of sexual vigor and muscular strength. It is probable that the extracts he prepared contained no sex hormones, and that the observed effects were imaginary (or the result of a placebo effect). Nevertheless, Brown-Séquard, confident after the experiment he had conducted in 1856 showing that removal of the adrenal glands caused the death of an animal, and of the hypothesis he had drawn from it that those glands contained substances indispensable to life, drew conclusions from the few observations made, and contributed to attracting great popular interest in endocrine glands and their secretions. His observations followed the principles of a therapeutic approach—organotherapy—that was fashionable at the time, which involved curing an illness using extracts from the organ whose dysfunction was responsible for the illness.[9]

In 1902, two English physiologists, William Bayliss (1860–1924) and Ernest Starling (1866–1927), observed that the duodenum stimulated the pancreas to secrete digestive enzymes even when the nerve connections between the two tissues had been cut. They showed that duodenum extracts produced the same effect, gave the active principle the name "secretin," and proposed a definition of what in 1905 would be called a "hormone": a substance produced in small quantities by an endocrine gland and circulated by the blood that acts on another organ.

In 1922, the Canadian physiologist Frederick Banting (1891–1941), in collaboration with Charles Best (1899–1978), purified insulin, secreted by the islets of Langerhans in the pancreas, the absence of which causes diabetes. In this case, too, many earlier observations—on the effects of the removal of the pancreas, or the results of work in anatomopathology—had suggested that the pancreas secreted an antidiabetic substance. But all attempts to purify it had failed. That failure turned out to be because insulin is a protein, and the pancreas also produces digestive enzymes—trypsin, chymotrypsin—that break down proteins. By ligating the excretory duct of the pancreas, Banting caused

9. Borrell, 1985.

the death of the cells producing the digestive enzymes, without altering the secretory function of the islets of Langerhans that produce insulin, thereby opening the door to the purification of insulin. The first clinical trials were carried out in the following months, and the Nobel Prize was awarded the next year.

Then it was the turn of the sex hormones to be purified and characterized. The chemists Adolf Windaus (1876–1959) and Leopold Ružička (1887–1976) would pave the way to their characterization, then to the synthesis of derivatives of cholesterol, including the sex hormones, but also vitamin D. In 1929, the German biochemist Adolf Butenandt (1903–1995), under the direction of Adolf Windaus, isolated estrone, and then testosterone in 1939. The study of sex hormones and of their role in reproduction enjoyed a great deal of support from very early on; there was hope for a better understanding, and thus for better control, of human reproduction. The pharmaceutical industry very quickly provided its assistance in producing the sex hormones and insulin. The Schering company collaborated very closely with Adolf Butenandt; producing hormones by extraction from animal tissue is a delicate process, and the biological activity of the products obtained must be monitored with great precision.[10]

After the war, the Schering company, in collaboration with the Flemish physiologist Ferdinand Peeters (1918–1998), developed a contraceptive pill. At the same time, Gregory Pincus (1903–1967) between 1956 and 1960 and following the work of Carl Djerassi (1923–2015), was carrying out his own trials on an oral contraceptive in Puerto Rico.

The characterization of hormones, and the complex biological regulation in which they participate, led the American physiologist Walter Cannon (1871–1945) to introduce in 1926 the

10. Gaudillière, 2005.

notion of homeostasis to account for the capacity of an organism to self-regulate.[11] This idea was the legacy of a long tradition going back to Hippocrates, by way of Claude Bernard; it also anticipated the general conception of such regulating phenomena that would be developed by cybernetics after World War II. The increasing tendency to explain biological phenomena in terms of regulation, which offered an organism the possibility to partially free itself from its external environment, also had a downside. It likely contributed to the difficulties biologists and doctors had at the beginning of the twentieth century in recognizing the existence of deficiency diseases. Indeed, how could it be explained that, if an organism maintains an unchanged internal environment, it nonetheless remains completely dependent on a few important trace substances derived from its external environment through food?[12]

The first attempts to isolate a hormone involved in the growth of plants—auxin—were made at the same time as those dealing with animal hormones, in the middle of the 1920s. The work, like that with animal hormones, elicited a great deal of interest, and many applications were envisioned.[13] But it would take until 1952 for the active substance to be definitively identified.

THE DISCOVERY AND CHARACTERIZATION OF NEUROTRANSMITTERS

The characterization of neurotransmitters, substances produced by neurons and capable of activating or inhibiting another neuron or muscle cell, is intimately linked to that of hormones.[14]

11. Wolfe et al., 2000.
12. Braun, 2011.
13. Rasmussen, 1999.
14. Dupont, 1999.

In 1856, the French physiologist Alfred Vulpian (1826–1887) described the presence in the adrenal glands of a substance that turns green in the presence of ferric chloride. In 1894 at University College London, the English physiologists George Oliver (1841–1915) and Edward Sharpey-Schafer (1850–1935) demonstrated that adrenal-gland extracts could shrink arteries and consequently cause an increase in arterial tension. In their publication they referred to the ductless secretory glands, such as the adrenals, as "endocrine" glands. Adrenaline was purified in 1901, and its structure established in 1907. In 1904, its presence was detected in the sympathetic nervous system, and it was suggested that it might act as a neurotransmitter, but the idea was not immediately accepted.

An experiment carried out by Otto Loewi (1873–1961) in Vienna in 1921 finally convinced scientists of the existence of neurotransmitters. He showed that an isolated frog heart placed in physiological liquid could be slowed through stimulation of the vagal nerve to which it remained connected, and that the liquid in which the heart was bathed in turn slowed the beating of a second heart. The substance responsible for the effect would be identified as acetylcholine, already known thanks to the work of the English pharmacologist Henry Dale (1875–1968).

In the 1940s, a very heated debate pitted Dale, who was convinced that the transmission of neural impulses between neurons was chemical (through neurotransmitters), against John Eccles (1903–1997), a proponent of direct electrical transmission. This resumed the debate, though in a different form, between Golgi and Cajal at the end of the nineteenth century. The reason for the debate was that chemical transmission had been demonstrated only in the case of the peripheral nervous system. What occurred at the level of the central nervous system remained a black box. A direct study of it was difficult, if not impossible. In addition, the neurotransmitters involved proved to be different from those active in the peripheral nervous system, and were substances so ordinary from a biochemical point of view, such as amino acids, that recognizing their role was more difficult.

The discovery of psychotropic substances in the 1950s would have a decisive, though indirect, role in the acceptance of the idea that the transmission of nervous signals at the central level is also chemical. How could scientists explain that chemical molecules could have such a great effect on the functioning of the central nervous system if that functioning didn't have a chemical base? Chemical transmission via neurotransmitters proved to be the most frequent form, but electrical transmission was also shown to exist between certain neurons. And contrary to what Dale had thought, it has been shown that a single neuron can produce several different neurotransmitters.

Adrenaline was distinguished from noradrenaline, other neurotransmitters—dopamine, glycine, gamma-aminobutyric acid—were described, and the pathways of their synthesis and breakdown were deciphered. In 1958 at Lund University, Arvid Carlsson (1923–2018) demonstrated that Parkinson's disease is caused by an absence of dopamine in a specific area of the brain called the substantia nigra, or black substance.

Like hormones, the characterized substances were very quickly used in medicine—following Carlsson's discovery, a precursor of dopamine, L-dopa, was proposed for the treatment of Parkinson's. As with hormones, there was a prejudice in favor of their therapeutic use because they were extracted from an organism—probably a hangover from organotherapy, but also a consequence of Claude Bernard's hypothesis that disease originates in a deviation from normality. The partial success achieved in the case of Parkinson's was not, however, repeated with other mental illnesses. The hope of being able to explain illnesses such as schizophrenia through the absence (or excess) of a single neurotransmitter quickly evaporated, and with it that of a simple pharmacological treatment of such illnesses.[15]

PROGRESS IN ELECTROPHYSIOLOGY

While chemical descriptions of the nervous system were being proposed, progress was also being made in its electrophysiological characterization, thanks to the development of microelectrodes, amplifiers, and oscilloscopes, enabling visualization of variations in electric potential and currents of low amplitude. In 1909, Louis Lapicque (1866–1952) introduced the notion of chronaxie—the time necessary to activate a neuron—which he interpreted as a sign of the process of integration carried out by the cell. Eccles precisely measured the pre- and postsynaptic potentials, and the positive or negative effects of the addition of neurotransmitters on those potentials. It was, moreover, the demonstration of the effects of postsynaptic inhibitors that convinced him of the reality of chemical transmission: it was impossible to explain the genesis of such effects if the transmission were electrical.

Working at the Plymouth Marine Laboratory in England, Alan Hodgkin (1914–1998) and Andrew Huxley (1917–2012) took advantage of the exceptional size of squid axons to place electrodes inside the nerve fiber and measure the ionic currents associated with the propagation of neural impulses. Their hypothesis was that these currents are due to the transitory opening of specific ion channels in the membrane, and in 1952 they

15. Kendler and Schaffner, 2011.

proposed a mathematical model accounting for their observations.

The connection between chemical and electrical approaches would also be made in the 1940s by the electrophysiologist Bernard Katz (1911–2003), who proposed an explanation of the quantum variations (leaps) in postsynaptic potential of neurons by the liberation of a fixed quantity of neurotransmitter—in the case he studied, acetylcholine. After the war, from observations of synapses under the electron microscope and from the discovery that neurotransmitters are stored in vesicles, confirmation would finally be achieved: the release of neurotransmitters corresponds to the exocytosis of those vesicles, which gives the process its quantum character.

INTEGRATIVE NEUROBIOLOGY

Neurophysiology didn't simply involve observations at the cellular level. Jean-Martin Charcot (1825–1893), a doctor at the Salpêtrière hospital in Paris, characterized many neurological diseases such as Parkinson's and, in 1868 (in collaboration with Alfred Vulpian), multiple sclerosis, which he distinguished from psychiatric illnesses. He described hysteria and the role that hypnosis could play in revealing it. His courses attracted audiences from all over Europe; however, Charcot himself recognized that they had turned into spectacles, and that the scientific value of what he was teaching had in part been diminished. His work had a great influence on Freud.

The work of Ivan Pavlov (1849–1936) was greatly influenced by that of Claude Bernard. Following Pavlov's work on the neural control of digestive secretions in dogs, which earned him the Nobel Prize, he demonstrated the importance of learning by association, known as conditioned reflexes. From his studies on the spinal cord, Charles Sherrington (1857–1952) showed the complexity of reflexes: a reflex movement demands simultaneous—and often occurring in opposite directions—control of the action of several muscles, leading Sherrington to assert the integrative function of the nervous system.[16]

Immunology, Microbiology, Virology, and Chemotherapy

CELLULAR IMMUNITY AND HUMORAL IMMUNITY

The first immunological mechanism to be described was phagocytosis, in 1882.[17] While he was working in Messina, the Russian biologist Élie Metchnikoff showed that introducing a needle into a starfish larva caused an affluence of cells called phagocytes, whose function was to rid organisms of foreign bodies. A confirmed Darwinian, Metchnikoff in 1883 inserted the phenomenon of phagocytosis into evolutionary history: phagocytes were derived from cells that, in simple multicellular organisms without digestive organs, provided nutrition. In more complex organisms, phagocytes became the principal agents of inflammatory phenomena that contribute to preserving the integrity of the organism and help to define its identity. They participate in embryonic development—they are responsible, for example, for the resorption of the tail of tadpoles—and in aging by attacking an organism's own constituents. Pasteur, though at first skeptical, nonetheless invited Metchnikoff to his institute in 1888. As we have seen, there Metchnikoff

16. Sherrington, 1906; Swazey, 1968.

17. Tauber, 2003. The information about immunology is drawn from Moulin (1991) and Brock (1999).

would develop a theory of aging through toxic bacteria, and attempted to perfect a lactic diet to prevent it.

The dominance of cellular immunology would be brief. As we have already seen, in 1890, Emil Behring and Shibasaburo Kitasato showed that the blood of animals into which they had injected toxins (diphtheria, tetanus) contained substances capable of neutralizing those toxins. The following year, Paul Ehrlich (1854–1915), who had participated in the production of a diphtheria antitoxin, though without his contribution being recognized, observed the same phenomenon after injecting plant toxins—such as ricin and abrin—and named those protective substances "antibodies." He conducted very precise assays of the activity of antibodies present in the serums. In 1897, he proposed the side-chain theory to explain this phenomenon of humoral immunity: cell protoplasm contains side chains to which toxins attach themselves. The organism reacts by overproducing these side chains to replace those that have been inactivated by toxins. The surplus passes into the blood, where it forms antibodies. The organism is protected because toxins attach to these circulating side chains and no longer react with the side chains of the protoplasm. Side chains contain two functional groups, which explains the phenomenon of precipitation observed between the serum and the toxins.

Jules Bordet (1870–1961), future founder and director of the Pasteur Institute of Brabant, demonstrated the presence in the blood of a factor, first called "alexine" and then "complement," which is rendered inactive through heat and is responsible for the bactericidal effect of antibodies.

So, two mechanisms were proposed to account for the phenomena of immunity.[18] Paul von Baumgarten (1848–1928)

18. Silverstein, 2003.

criticized Metchnikoff for not explaining cellular immunity in physicochemical terms, as the theory of antibodies did.

Almroth Wright brought the two mechanisms together in 1904 by demonstrating that the attachment of antibodies to bacteria facilitated their phagocytosis, a phenomenon to which he gave the name "opsonization." In 1908, the Nobel committee granted its prize to both Ehrlich and Metchnikoff. Nevertheless, work on antibodies (humoral immunity) developed much more quickly than did that on cellular immunity, and it wasn't until the second half of the twentieth century that work on cellular immunity would regain ground.

More than the medical applications—the development of serotherapy, as we have seen—it was the possibility of a very precise characterization of the interactions between antibodies and what would be called antigens that inspired physicochemists such as the Swedish Svante Arrhenius (1859–1927) to study them. In 1907, he devoted a book to immunochemistry. Michael Heidelberger (1888–1991) and Oswald Avery (1877–1955) characterized the antibodies targeted against pneumococcus and demonstrated that they were proteins. Karl Landsteiner (1868–1943) would be the mastermind in the development of this immunochemical approach. In 1900, he characterized the ABO blood groups by studying reactions of crossed immunoprecipitation between red cells and serums from different individuals. After centuries of failed attempts at blood transfusion, the description of blood groups enabled the process to be developed. In 1940, with Alexander Wiener (1907–1976), Landsteiner discovered the rhesus system, responsible for fetal-maternal incompatibilities.

Starting in 1917, Landsteiner carried out a systematic program to study antigen-antibody reactions by modifying antigen molecules in a very targeted and selective way. He showed the extraordinary specificity of the antigen-antibody reaction in his work *Specificity of Serological Reactions*, published in German in 1931 and translated into English in 1945. This, in turn, led to the ability to distinguish between molecules with similar structures and thus show the specificity of the components of living things, and in particular proteins. A protein, such as trypsin, extracted from one species, was revealed to be different from the homologous protein extracted from another species. The importance of the notion of the specificity of living organisms grew at the same time as techniques of immunochemistry were being developed, owing to the revelatory work of Michael Heidelberger and Elvin Kabat (1914–2000), in particular. In the 1940s, in a research movement led by the American chemist Linus Pauling (1901–1994), this notion of specificity found its precise chemical counterpart in stereospecificity. Pauling proposed that an antibody's affinity for an antigen, like that of an enzyme for its substrate or of a receptor for a hormone or neurotransmitter, was due to the formation of a set of "weak" bonds (hydrogen bonds, ionic bonds) between the two molecules. As the strength of these bonds is highly dependent on interatomic distance, binding occurs only if there is a perfect structural complementarity between the two molecules, such as that of a key and a lock, to borrow the image proposed earlier by Emil Fischer.

BACTERIAL MICROBIOLOGY AND THE DISCOVERY OF VIRUSES

At the same time as immunochemistry was being developed, the characterization of new microbes and the development of vaccines and serums continued.[19] In 1906, Bordet showed that

19. Berche, 2007.

whooping cough was due to a bacterium, since named *Bordetella pertussis*. After the fruitless attempts by Koch, an effective vaccine against tuberculosis was finally developed at the Pasteur Institute in Lille, and then at the one in Paris, by Calmette and Camille Guérin (1872–1961).

Immediately after the discovery of the agent of tuberculosis, *Mycobacterium tuberculosis*, and of bovine tuberculosis, *M. bovis*, there had been attempts to apply the principle of vaccination directly; but the result was catastrophic, as the bovine bacterium proved to be as pathogenic for human beings as *M. tuberculosis*. Nevertheless, Calmette and Guérin worked with the bovine bacterium, which they attenuated using the Pasteurian method of culturing in vitro for more than 11 years. The first trials were carried out on human beings in 1921. The BCG vaccine—named in homage to its creators (bacillus Calmette-Guérin)—aroused a certain suspicion. This was reinforced in 1930 by the "Lübeck affair," in which 72 out of 251 newborn babies died, and 131 developed tuberculosis, after being accidentally infected by an active strain of the bacterium.

Gaston Ramon (1886–1963) improved the production of vaccines against diphtheria and other microbes by rendering them or their toxins inactive through formalin.

Despite these successes, the debate over the specificity of pathogenic microorganisms, begun by Koch and Nägeli, continued. Some scientists proposed cyclogenic theories for bacteria, according to which bacteria would be converted into organisms of different forms in the course of complex life cycles.

Microbiology was transformed with the emergence of a new type of pathogenic agent: the virus.[20] The Berkefeld filter, developed in Germany, and in France a filter enabling the reten-

20. Hughes, 1977.

tion of pathogenic microbes (invented by Charles Chamberland [1851–1908], a collaborator of Pasteur), were the tools for their discovery. In 1892, the Russian botanist Dmitri Ivanovsky (1864–1920) had described a pathogenic microorganism of tobacco that was not retained by filters (though he remained vague about its nature), a result confirmed by Martinus Beijerinck (1851–1931) in Delft in 1898. Beijerinck proposed the already widely used name of "virus" ("poison" in Latin) to describe this new type of pathogenic agent. The same year, the German bacteriologists Friedrich Loeffler (1852–1915) and Paul Frosch (1860–1928) demonstrated that foot-and-mouth disease was also due to the action of a virus.

Gradually, over the years that followed, many other viruses were discovered. In 1909, Landsteiner and Constantin Levaditi (1874–1953) revealed the poliomyelitis virus. A second criterion of characterization was added to viruses' small size: the impossibility of making them multiply in a liquid environment. Abandoning the initial idea that they needed a particular substance to replicate, the bacteriologist Thomas Rivers (1888–1962) in 1926 adopted the hypothesis that they are obligate parasites. The development of the electron microscope just before World War II, and the biochemical characterization of viruses, would contribute further criteria to distinguish them from other pathogenic agents.

The impossibility of cultivating viruses in vitro made the development of vaccines problematic. In 1931, the American pathologist Ernest Goodpasture (1886–1960) demonstrated that it was possible to make viruses multiply in fertilized chicken eggs, and this opened the path to the mass production of viruses, and thus to the production of vaccines. After World War II, culturing in isolated cells would progressively replace culturing in chicken embryos. In 1952, the development of a vaccine against poliomyelitis by Jonas Salk (1914–1995), and in 1957 of an attenuated, orally administered vaccine by Albert Sabin (1906–1993), would represent major advances in the fight against viral diseases.[21]

In 1915 and 1917, the English bacteriologist Frederick Twort (1877–1950) and the French-Canadian Félix d'Hérelle (1873–1949), working at the Pasteur Institute in Paris, independently discovered the existence of viruses that infected bacteria, called bacteriophages. It was d'Hérelle who put these minuscule objects at the center of biology.[22] Unlike biologists such as Bordet who believed them to be an endogenous product of bacteria, d'Hérelle was determined to show that bacteriophages were independent organisms. Their small size and their rapid reproduction made them a model system for the study of the fundamental properties of life. D'Hérelle thought that organisms similar to bacteriophages represented the first living beings, whose later forms appeared through a process of accretion. He was also convinced that it would be possible to use bacteriophages as therapeutic agents. Early observations were made in 1896 by the British bacteriologist and chemist Ernest Hankin (1865–1939), and published in the *Annales de l'Institut Pasteur*. Hankin noted a bactericidal action of water from the Yamuna and Ganges rivers, even after filtration—without, however, being able to interpret his observations. D'Hérelle attempted to demonstrate the effectiveness of bacteriophages against the plague and cholera in India. A student of his from Soviet Georgia developed the therapeutic use of d'Hérelle's findings in his own country, which was then extended throughout the entire Soviet Union. The use of bacteriophages would be almost completely abandoned after the

21. Oshinsky, 2005.
22. Summers, 1999.

development of antibiotics, however, and their therapeutic effectiveness in retrospect remains difficult to assess.

The great flu pandemic of 1918–1919 reveals the limits of the knowledge of microbiologists at the time, because the characterization of viruses had not yet led to new therapeutic procedures. While microbiology was going through a difficult period in the 1920s and1930s, the its researchers were romanticized in two best-selling books. Sinclair Lewis's *Arrowsmith*, published in 1925, describes an idealistic microbiologist (who was identified with Félix d'Hérelle) who uses bacteriophages to fight against a plague epidemic. And in 1926, the microbiologist Paul de Kruif (1890–1971) published *Microbe Hunters*, a book that was also very popular. Because of their small size and their simplicity, in the mid-twentieth century viruses and bacteriophages would also be highly valued as models by molecular biologists seeking "the secrets of life."[23]

Microbiology also has another side, one that is less medical and draws less media attention, but which is no less important: its contribution to the study of metabolism, and the use of microorganisms in biotechnology. It was the scientific school of Albert Jan Kluyver (1888–1956)[24] in Delft that characterized bacterial metabolism. That work revealed the identity of the principal metabolic pathways in very different organisms, which would lead to a unified view of the living world, and which would support the development of a modern evolutionary synthesis. In 1926, in his book *Unity in Biochemistry*, Kluyver describes this continuity "from elephant to bacterium," a phrase that would later be attributed erroneously to Jacques Monod (1910–1976).[25]

23. Creager, 2002.

24. Singleton and Singleton, 2017.

25. Brock, 1999.

Microorganisms were also used to manufacture chemical molecules intended for industry. Chaim Weizmann (1874–1952), future president of Israel, and Auguste Fernbach (1860–1939) of the Pasteur Institute in Paris used bacteria to break down starch into acetone and butanol. An initial plan to use these two molecules to manufacture rubber did not come to fruition, but the fermentation process would be used again during World War I to manufacture explosives.

THE GROWING IMPORTANCE OF CHEMOTHERAPY

The beginnings of chemotherapy—use of chemical agents to treat disease—are less spectacular than those of microbiology. Paul Ehrlich defined its principles and developed the first methods of systematic screening: his early work consisted of testing the affinities of different dyes produced by the German chemical industry for cellular components. Success in obtaining dyes specific to certain cell types or cell structures convinced him that it should be possible to find molecules, which he called "magic bullets," capable of attaching to microorganisms and killing them without those substances affecting the host organism. He demonstrated that methylene blue is active against the agent of malaria, but that the side effects proved too harmful to enable its therapeutic use. Collaborating with the German company Hoechst, he characterized a molecule—molecule 418—active against sleeping sickness. In 1910, he described salvarsan, active against syphilis. The failure of Koch's use of tuberculin as an antituberculosis medicine made it more difficult for people to accept these new therapeutic agents. In addition, their use sometimes led to serious therapeutic accidents. Tryparsamide, active against trypanosomes, the causal agents of sleeping sickness, was used widely in the 1920s by Eugène Jamot (1879–1937) in the Cameroon and the Upper Volta. A dosage error by one of his staff caused blindness in 700 of the patients who had been treated. After World War II, another medication used for sleeping sickness, lomidine (pentamidine), again caused serious complications.[26]

Efforts to find chemical molecules active against microorganisms would continue unabated. In 1932, Gerhard Domagk (1895–1964), working at Bayer, described the antibacterial action, in particular against streptococci, of a sulfonamide, prontosil. Daniel Bovet (1907–1992), Federico Nitti (1903–1947), Jacques Tréfouël (1897–1977), and Thérèse Tréfouël (1892–1978), working in the laboratory of Ernest Fourneau (1872–1949) at the Pasteur Institute in Paris, showed that the active part of the molecule was sulfanilamide, which they immediately used in therapeutic trials on human beings. Sulfonamides were widely used by the American army during World War II before they were replaced by antibiotics.

The idea that organisms could prevent the growth of and/or kill other organisms was widely accepted. The antibacterial effect of a *Penicillium* mold had been observed several times before Alexander Fleming (1881–1955) in 1928 gave a precise description of it. Observing that the molds acted on gram-positive and not on gram-negative bacteria, he proposed using them to purify certain bacteria, or as antiseptics.[27] In 1939, the Australian Howard Florey (1898–1968) attempted to use the substance produced by *Penicillium* as a therapeutic agent. The trials were conclusive, but limited by the difficulty of isolating the active agent. Two chemists, the German-born Ernst Chain (1906–1979) and

26. Lachenal, 2014.
27. Chen, 1996.

Norman Heatley (1911–2004) at Oxford University, changed the culture conditions to increase production, developed a protocol for isolation of the active substance, and adapted those conditions to mass production.[28] Over two million doses of penicillin would be ready for the Normandy invasion. René Dubos (1901–1982), working at the Rockefeller Institute in New York, characterized two other antibiotics, gramicidin in 1938 and tyrothricin in 1941–1943. In 1943, after a systematic study of active substances produced by microorganisms in the soil, Selman Waksman (1888–1973) introduced the term "antibiotic" to describe them, and his student Albert Schatz (1920–2005) discovered streptomycin, the primary active compound against tuberculosis.

Developmental Biology and Cellular Biology

THE "MECHANICS OF DEVELOPMENT"

The recapitulation model proposed by Haeckel elevated work in embryology—the precise description of embryonic development often being the best means to classify organisms—but simultaneously forced that discipline to abandon any explanations that were unique to it, as fewer and fewer people were interested in the mechanisms of development.[29] Wilhelm Roux, a student of Haeckel, would retain from his teacher the importance of Darwinism and of competition, and he even introduced the idea of competition among the parts of an organism. But he refused to admit that the recapitulation of evolution

28. Lax, 2004.

29. Information in the section "Developmental Biology and Cellular Biology" is drawn from Horder et al. (1986), Gilbert (1991), Nyhart (1995), and Maienschein (2014).

was an explanation for embryonic development. He called for the creation of a "developmental mechanics" to describe the "proximal causes" of embryologic phenomena. Mechanisms of development could be proven only by direct experimentation. Wilhelm His (1831–1904), a student of Virchow and professor of anatomy in Basel, had already criticized the theory of recapitulation and called for a direct study of mechanisms of development.

Nor was manipulating embryonic development to reveal its mechanisms a new idea. Étienne Geoffroy Saint-Hilaire, and later his son Isidore (1805–1861), had developed an experimental teratology. Being interested, as Wilhelm Roux was, in the first phases of embryonic development, Laurent Chabry (1855–1894) destroyed certain cells of an early embryo of an ascidian and studied the consequences of this on embryonic development.[30]

However, it was the contradictory experiments of Roux and Hans Driesch (1867–1941) that energized this new approach.[31] In 1888, Roux, using a heated needle, had destroyed one of the first two cells from the division of the egg of a frog, and showed that, under those conditions, he obtained only a half-embryo. He suggested the "mosaic" development model: each part of the egg contributes to the formation of a different part of the organism. The proposed explanation was that germplasm, whose characteristics were described by Weismann, was distributed unequally among cells during development. Driesch, working at the zoological station in Naples, performed an analogous experiment on a sea urchin embryo . . . and obtained different results. When he separated the first two cells from the division of the egg through agitation, or by placing them in pure

30. J.-L. Fischer, 1991.

31. For Driesch, see Allen (2008).

water, each of them produced an entire organism. There was thus a process of regulation that intervened during development, which depended on contacts between cells. Roux attempted to disqualify Driesch's experiment by hypothesizing that Driesch was observing an abnormal phenomenon of rapid regeneration. In 1901, Hans Spemann (1869–1941) confirmed Driesch's results; in retrospect, Roux's results were explained by the fact that killing one cell prevented the development of the other cell. Driesch showed that each cell remained "totipotent"—that is, capable of generating an entire organism—up to the stage when the organism consists of four cells.

Driesch then turned to philosophy. In his opinion, the results he had obtained showed the presence in the embryo of a "plan," of a principle guiding development, which he called "entelechy." His reintroduction of a form of vitalism and his interest in parapsychology gradually marginalized him.

In 1912, by modifying the composition of the environment in which embryonic development took place, Jacques Loeb (1859–1924) at the Rockefeller Institute in New York achieved an artificial parthenogenesis in a sea urchin—that is, the starting of embryonic development without fertilization of the oocyte by a spermatozoon. His experiment had a considerable impact. It revealed the knowledge that had been acquired by embryologists in the manipulation of embryonic development, and was consistent with the concept that Jacques Loeb proposed in his 1912 book, *The Mechanistic Conception of Life*.[32]

At the same time, several biologists were attempting to create a living organism in vitro, in Spain, in Mexico, and in France in the lab of Stéphane Leduc (1853–1939).[33] If the mechanistic

32. Pauly, 1987.

33. Keller, 2002.

conception of life and evolutionary theory—which suggests a "natural" origin of living forms—rendered such projects in synthetic biology reasonable, the state of knowledge of the fundamental mechanisms of life limited their chances of success. To those involved, being able to reproduce certain forms of life—which Maupertuis had already done—seemed enough to assert that they had created life in vitro. That research would fade away without leaving any direct legacy.

THE CELL LINEAGE

Opting for a completely different approach, several biologists described precisely the first stages in the division of an egg, thereby establishing what is called a "cell lineage."[34] Their results also challenged Haeckel's law of recapitulation. In 1892, Theodor Boveri described the development of *Ascaris*. He showed that during its development, chromatic reduction—loss of chromosomes—occurs. In 1878 in the United States, Charles Whitman (1842–1910) performed an analogous study on the leech *Clepsine*. In the marine biology lab that Whitman founded in Woods Hole near Boston in 1888, Edmund Wilson (1856–1939) and Edwin Conklin (1863–1952) performed similar studies in the annelid worm *Nereis* and the gastropod *Crepidula*. The goal was to determine whether embryonic development perfectly recapitulates the evolutionary past, or if it is the result of an adaptation by the embryo to the specific conditions surrounding its development. The results were ambiguous: without invalidating a relationship between embryonic development and evolutionary history, they revealed differences in the process of development that only an experimental approach could potentially account for.

34. Maienschein, 1978.

RESEARCH ON INDUCERS

The most remarkable results of experimental embryology were obtained by Spemann and his doctoral student Hilde Mangold (1898–1924);[35] they were published the year Hilde died. The grafting of a small section of a salamander embryo (the dorsal lip) induced in the receiving embryo the formation of a new organism, thereby creating two conjoined twins. Hans Spemann described other inducive effects that occurred during embryogenesis, although they were less spectacular than those of the structure he called the "organizing center," or, more simply, the "organizer." Salamanders were chosen for these experiments because different varieties had embryos with different pigmentation, making it possible, after the graft, to distinguish the tissue of the receiving embryo from the grafted tissue.

Spemann was criticized, especially by Thomas Hunt Morgan (1866–1945) and other American geneticists, for having totally ignored the existence of genes in the interpretation of his results, and for proposing a view of development described as "vitalist."

In the 1930s in England, Joseph Needham (1900–1995)—who became famous after World War II for his work on the history of science in China—and Conrad Waddington (1905–1975), in contrast to the holistic approach of Spemann, attempted to characterize the chemical nature of the inducive substance present in the organizer. The results were disappointing and disturbing: not only did they fail to characterize it, but many simple organic molecules proved capable of mimicking the action of the organizer.

Spemann's experimental approach was adopted by several embryologists, including Viktor Hamburger (1900–2001) and

35. For Spemann, see Horder and Weindling (1986).

the Austrian Paul Weiss (1898–1989) who, through analogy with physics models, developed the notion of a morphogenetic field to account for the effects of embryonic induction.[36] After he emigrated to the United States, Hamburger demonstrated the existence of neuron growth factors, which Rita Levi-Montalcini (1909–2012) would purify.

THE FIRST STEPS IN *EX VIVO* CULTURING

To study the development of nerve cells, the American embryologist Ross Harrison (1870–1959), between 1907 and 1910 at Yale University developed a system of culturing explants of embryos in lymphatic liquid.[37] He could thereby observe the growth of neurons and the establishment of neural connections.

Keeping organs alive outside an organism wasn't new. In the nineteenth century, the biologist Julius Sachs had isolated and kept alive fragments of plants. The hearts whose slowing Otto Loewi observed in 1921 were kept alive in Ringer's solution for several hours. What was original in Harrison's work, and enabled by the specific medium he had used, was the observation of a process of embryonic development in an isolated animal tissue.

Alexis Carrel (1873–1944) was a French-American researcher known for his work on the suturing of blood vessels and the development—thanks to the patronage of Charles Lindbergh—of an apparatus for extracorporeal blood circulation (in the goal of developing an artificial heart). Carrel claimed to have kept a frog heart alive for several decades, and, although contested today, his results caused considerable media interest.

36. For Hamburger, see Allen (2004).

37. Witkowski, 1986; Slack, 2003.

Culturing isolated cells of an organism (cell culture, rather than tissue culture) was achieved only after World War II, under the impetus of John F. Enders (1897–1985) and thanks to the advent of antibiotics, which could protect cells from contamination for a very long time.[38]

It wasn't until the 1960s that plants also benefited from techniques for culturing isolated cells. The need to break down the cellulose wall that surrounds plant cells—a technique that was difficult to develop—and the fear that the process would make cells lose their characteristics, in part explains the delay. Once those obstacles were overcome, cell culture contributed greatly to the development of plant biotechnology.

The Rediscovery of Mendel's Laws, and the Rise of Genetics

THE FIRST STEPS IN GENETICS

At the beginning of the year 1900, in the *Comptes rendus de l'Académie des sciences de Paris*, Hugo de Vries published the results of his experiments in the hybridization of peas.[39] The proportions in which the first-generation hybrids produced the hybrid type or the parental types corresponded to what Mendel had described 35 years earlier. Not long afterward, the German botanist Carl Correns (1864–1933) published results similar to those of de Vries, referring back to Mendel's results in his report. He criticized de Vries for not mentioning Mendel, and de Vries would do so in his subsequent publications. The same year, the Austrian botanist Erich von Tschermak (1871–1962)

38. Landecker, 2007.

39. Information in "The Rediscovery of Mendel's Laws, and the Rise of Genetics" section is drawn from Sturtevant (1965).

would also announce analogous results; he already knew of Mendel's work, although he hadn't perfectly understood it.

The focus of de Vries's and Correns's work was, however, quite different from Mendel's. For them, it was a matter of studying the transmission to descendants of the mutations revealed by de Vries.

The impact of this work was enormous. In Nancy, France, the zoologist Lucien Cuénot (1866–1951) immediately began experiments crossing mice with different colored coats, and in 1902 he showed that the laws established by Mendel for peas could be applied to animals. In the following years, he revealed the existence of lethal forms of what a few years later would be called "genes" (1909), found evidence for epistasis—that is, interactions among genes—and showed that a genetic variation could have multiple effects: the phenomenon of pleiotropy.

In England, William Bateson (1861–1926) was the champion of what in 1905 he called "genetics."[40] In 1880, he was interested in variations likely to play a role in the evolution of organisms, and in 1894 described homeotic variations—those that produce a repetition of body parts—in his book *Materials for the Study of Variation*.[41] Like de Vries, he was convinced that evolution could be discontinuous, could occur in leaps, although he didn't launch a frontal attack on Darwinian theory. In 1899 he had begun crossing experiments and, after the rediscovery of Mendel's laws, like Cuénot he very rapidly showed that those laws could also be applied to animals. He introduced the term "alleles" to describe the different forms of a gene, and "homozygote" and "heterozygote" to distinguish organisms having (or not having) two identical copies of the same gene. He

40. Darden, 1977; Cock and Forsdyke, 2008; Peterson, 2008.
41. Bateson, 1894.

demonstrated the superiority of the new theory compared with biometry: whereas biometry was focused on the past and explained characteristics of an organism by those of its ancestors, genetics looked to the future and enabled scientists to foresee the effects of crosses. Shortly after the rediscovery of Mendel's laws, Bateson, with Edith Rebecca Saunders (1865–1945) and Reginald Punnett (1875–1967), noted that not all characteristics are transmitted completely independently, but that some seem to be transmitted together, a phenomenon to which he gave the name "genetic linkage."

In 1909, the Danish geneticist Wilhelm Johannsen (1857–1927) introduced the term "gene," and the distinction between the genotype and the phenotype—that is, between the characteristics of genes and those of the organism. The work he carried out in Lund, Sweden, consisted of producing pure lineages of beans and studying variations among different plants; to use the terms he himself introduced, his goal was to determine the one or more phenotypes associated with a given genotype.

This work was important because it debunked the hypotheses of biometricians. As we have seen, biometricians had proposed a law of regression according to which the crossing of two individuals whose characteristics were identical to each other's but different from the population mean, would produce descendants whose characteristics were centered around a value located between those of the parents and those of the population mean. Johannsen showed that this principle of regression didn't apply to the pure lineages of the beans on which he was working. Regardless of the parental characteristics for the trait being studied, the descendants were distributed around the general mean. Thus, Johannsen showed that biometricians had confused two different types of variation: variation due to the environment, and variation of the genotype. The law of regression was explained by a combination of these two phenomena in populations with different genotypes: small phenotypical variations connected to the environment were not transmitted, whereas those owing to the genotype were.

Genetics was in a paradoxical situation. On the one hand, it was very quickly accepted by many biologists and was used as a viable method to predict the results of crosses. In the United States, the first chairs in genetics appeared in recently established universities with agriculture programs. In France, the Vilmorin seed company was one of the organizers of the first genetics conferences, held in Paris.[42] However, geneticists didn't agree on the nature of the gene. Some were content to use the laws of genetics without questioning the nature of the "factors" that are transmitted from generation to generation. Others, like Bateson himself, thought that the gene was not an object, but something else—a vibration, a wave, something dynamic. This uncertainty about the foundations of genetics can be seen in the reticence of several of the first geneticists to associate genes with chromosomes, as Theodor Boveri and Walter Sutton (1877–1916) had proposed doing in 1902, pointing out the parallel between the transmission of genes following Mendel's laws, and that of chromosomes through the generations.

THE MORGAN SCHOOL OF GENETICS

The results of Thomas Hunt Morgan in part, but only in part, caused the scales to tip in favor of a material conception of the gene.[43] They also provided tools enabling the order of genes

42. Gayon and Zallen, 1998.
43. Allen, 1978.

on the chromosome to be determined. Morgan was an embryologist whose early work, carried out at the Woods Hole station and following the prescriptions of Haeckel, was studying the embryonic development of the sea spider in order to place it on the tree of life. In 1894, at the zoological station in Naples—thanks to Driesch—he marveled upon discovering the new mechanics of development. Back in the United States, Morgan attempted, with mixed results, to use the new experimental approach to better understand the phenomenon of regeneration.[44] He showed that regeneration was not a distinct adaptive process selected by evolution, but a phenomenon strictly associated with the process of embryonic development. Influenced by Driesch, he proposed a holistic conception of regeneration, quite different from the reductionism he would demonstrate in his activities as a geneticist. At the same time, he immersed himself in the mechanisms of evolution. Convinced by de Vries that evolution occurred through mutations, in 1908 he chose a model organism that reproduced quickly and could thus easily demonstrate such mutations. He chose the fruit fly, which had already been used by the American geneticist William Castle (1867–1962). The first two years would be disappointing: no large-scale mutation was apparent. However, a small variation in the color of the flies' eyes (some had turned white) caught Morgan's attention, because its transmission seemed to depend on the gender of the flies. Connecting this observation to the chromosomal theory of gender proposed by Wilson in 1905, he interpreted the results with his newly created hypothesis that the genic form responsible for the white color of the eye was carried by one of the chromosomes that determine gender.

44. Sunderland, 2010.

The chromosomal positioning of genes was confirmed by his own observations of genetic linkage and by the earlier observations of Bateson. Genetic linkage was easily explained by the assumption that the two linked genes were carried by the same chromosome: the specific forms of these genes present on the same chromosome were transmitted together and not independently of each other as Mendel had asserted.

But genetic linkage was not perfectly stable, which seemed to contradict the hypothesis. Once again, Morgan connected these observations to those made a little earlier, in 1909, by the biologist Frans Janssens (1865–1924) from Leuven: during meiosis—that is, cell division in the formation of reproductive cells, spermatozoa and oocytes—the two chromosomes of the same pair combine. This pairing can lead to breakage followed by reconnection, thereby enabling an exchange (called recombination) of alleles carried by the two chromosomes of the same pair.

Alfred Sturtevant (1891–1970) proposed the simplifying hypothesis that the probability of breakage is constant all along chromosomes: meaning that two genes will have a greater chance of being separated during meiosis the farther apart they are on the chromosome. Frequencies were thus replaced by a map; the first genetic map was constructed by Sturtevant in 1913.

Geneticists henceforth had a well-established research program: characterize mutations and position them on increasingly complex genetic maps. The fly room at Columbia University in New York where Morgan, Sturtevant, Hermann J. Muller (1890–1967), and Calvin B. Bridges (1889–1938) worked was the place where the chromosomal theory of heredity was born.[45] The study of other organisms would quickly follow: corn, different

45. Kohler, 1994.

mammals—thanks to the work of Castle[46]—and human beings, despite the impossibility of carrying out crosses in them. In 1927 Muller showed that X-rays have a mutagenic effect. Geneticists could henceforth "force" the appearance of mutations and no longer had to wait passively for them to appear.

Thanks to the work of Cyril Darlington (1903–1981) and Barbara McClintock (1902–1992) on corn, the morphological (cytological) study of chromosomes joined the observations of geneticists. At the beginning of the 1930s, scientists proved the existence of giant chromosomes (called polytene chromosomes because they resulted from chromosomal duplication without separation of the daughter chromosomes or cell division) in the salivary glands of fruit flies. These giant chromosomes showed an alternating pattern of light and dark bands that was very similar, sometimes identical, from one fly to another. By matching deletions (absences) and rearrangements on genetic maps with the modification of chromosomal bands visible under the microscope, geneticists could superimpose the morphological maps and the genetic maps.

Genetics didn't develop at the same rate in every country. In Germany, unlike the United States, genetics didn't become an independent discipline. Work in genetics for the most part was associated with work in physiology or morphology. Thus, Alfred Kühn (1885–1968) with his student Ernst Caspari (1909–1988), working in Göttingen and then in Berlin, attempted to connect the physiological mechanisms of development to the action of genes by using a model organism—the flour mite *Ephestia*. In 1924, he isolated mutations altering the shape and color of the wings, then others modifying the color of the eyes. This work, like that of Boris Ephrussi (1901–1979) and George

46. Snell and Reed, 1993.

Beadle (1903–1989), which we will discuss below, showed genetic control in the production of enzymes, but didn't reveal anything about the genetic mechanisms of development.

Correns, but above all Erwin Baur (1875–1933) and Otto Renner (1883–1960), demonstrated and characterized phenomena of cytoplasmic inheritance, independent of the chromosomes of the cell nucleus.[47] This inheritance is connected to plastids (chloroplasts, organelles in cells responsible for capturing light energy). Plastids of paternal origin were found to be broken down after fertilization. These phenomena were completely ignored by the English and American schools of genetics.

In France, genetics remained a marginal science. Lucien Cuénot interrupted his work after World War I, both because of the loss of his samples during the war, and because he was discouraged by how difficult it was to have his results accepted. Genetics was indirectly handed down from Weismann, and thus viewed with suspicion by French biologists, most of whom were neo-Lamarckians. Its results weren't ignored, but the characteristics whose transmission geneticists were studying were considered secondary. With a few exceptions, French breeders and seed producers also saw genetics as only a marginal science, preferring to discuss their products in terms of varieties rather than genes.[48]

THE TEMPTATION OF EUGENICS

The relationships between eugenicists and geneticists were complex.[49] Inspired by Galton, eugenicists adhered primarily

47. For Correns, see Rheinberger (2000a).

48. Bonneuil, 2006.

49. Information in "The Temptation of Eugenics" subsection is drawn from Kevles (1998) and Carlson (2001).

to the biometrical model. When, following the work of Johannsen, biometrics lost its importance as a theory of heredity and its quantitative methods were gradually integrated into genetics, eugenicists and geneticists drew closer. The eugenics societies that formed were in most cases societies for both eugenics and genetics. Everyone found something of interest in them: eugenicists found scientific security, and geneticists resources and increased visibility. The attitude of geneticists was varied. Some were convinced of the need to implement a eugenics policy, whereas others, like Morgan, were much more guarded vis-à-vis the simplistic view of eugenicists and the often racist connotations of the measures they wanted to put into place. Genetics was used in eugenics policies, as in Nazi Germany, with the collaboration of many geneticists.

The Colchester study, conducted in 1938 by Lionel S. Penrose (1898–1972), represented the first serious inquiry into the genetic determinism of mental retardation, which was the preoccupation of eugenicists. In their opinion, the very existence of mental defects demanded the implementation of a policy of forced sterilization. The study revealed the diversity of cases that fell under the category of "mental retardation," and thus spelled the end of the simplistic models proposed by eugenics.

The Rise of Molecular Biology

THE DEVELOPMENT OF NEW TECHNOLOGIES

It is difficult to define molecular biology, and to demarcate the borders between it and biochemistry and genetics.[50] It was

50. Information in "The Rise of Molecular Biology" section is drawn from Olby (1974), Judson (1996), Kay (1993), Fruton (1999), and Morange (2020).

born out of the encounter between those two disciplines, and results in molecular biology established bridges between them. Molecular biology describes the relationships that exist between genes (the object of study of geneticists) and proteins (which have a central role in biochemistry). The characterization of these bonds necessitates the characterization of the chemical structure of genes (DNA) and that of proteins.

The rise of molecular biology rested on the development of the techniques that enabled isolation of macromolecules and the study of their structure. The first was ultracentrifugation, developed by Theodor Svedberg (1884–1971) in Uppsala, Sweden, in the 1920s. The outcome of Svedberg's first experiments allowed him to demonstrate that proteins, such as hemoglobin, have a precise molecular mass, and confirmed macromolecular theory at the expense of colloidal theory.[51]

Electrophoresis is probably the technique most emblematic of molecular biology. It consists of the separation of macromolecules by their differential migration in an electrical field.[52] It was developed by one of Svedberg's students, Arne Tiselius (1902–1971), in Uppsala between 1925 and 1935. The preparation in the 1950s and 1960s of new supports for electrophoresis, such as the well-known acrylamide gels, and new methods for preparing samples—with the use of agents such as SDS (sodium dodecyl sulfate) for denaturing proteins—constituted the major stages in the development of molecular biology. The development of agarose gels in the 1970s played a role for nucleic acids, and for the development of genetic engineering, comparable to that of acrylamide gels in the study of proteins.

51. Deichmann, 2007.
52. Chiang, 2009.

Other techniques, developed around the same time, were also used in the exploration of this unknown world that existed between the molecules of the organic chemist and the subcellular structures barely discernable under an optical microscope. The electron microscope, which uses electrons instead of visible light, was developed by the German physicist Ernst Ruska (1906–1988) in collaboration with the Siemens company at the beginning of the 1930s. Viruses were the first biological entities to be studied. Ultraviolet spectroscopy—that is, the study of the absorption of ultraviolet light by biological molecules—was developed in the same period: it enabled quantification of DNA and proteins. The diffraction of X-rays played an essential role in the determination of the structure of DNA and that of proteins. The principles of this process were established by the German physicist Max von Laue (1879–1960) and the English physicist William Lawrence Bragg (1890–1971) between 1910 and 1920. In 1934, the Irish crystallographer John Bernal (1901–1971) obtained the first precise X-ray diffraction images of the crystals of a protein, pepsin, by keeping it in its crystallization liquid. The impact of that observation, published with Dorothy Hodgkin (1910–1994),[53] was enormous: not that the image immediately revealed anything of the protein's structure, but because the precision of the image obtained confirmed that proteins have a perfectly defined spatial structure.

Two other contributions from chemists played an essential role in the determination of macromolecular structures. The development of chromatography enabled scientists to do without the traditional techniques of precipitation used by organic chemists to separate molecules, techniques that often denatured macromolecules. Furthermore, the characteristics of the

53. Howard, 2003.

chemical bonds, the distances between atoms, the sizes of the angles between two bonds formed by the same atom, were determined with precision, both through direct X-ray diffraction measurements of small molecules, and by the application of quantum theory to chemical bonding. Linus Pauling's book, *The Nature of the Chemical Bond*, published in 1939, allowed organic chemists to use the results of those theoretical and experimental studies. Pauling also contributed to the characterization of weak bonds and proposed a model to explain the catalytic action of enzymes (proteins): enzymes bind to with a great affinity—and thereby stabilize—the transition state of substrates during the reactions that they catalyze.[54] They thus favor the conversion of the substrate into the product.

The development of new technologies such as ultracentrifugation and electrophoresis demanded a lot of effort: the small size (and simplicity) of the apparatuses used today for electrophoresis are a far cry from the bulk and complexity of the first machines, which took up entire rooms and demanded the permanent presence of many engineers. The development of these first machines, and their acquisition by laboratories, was overseen by the Rockefeller Foundation, under the guidance of its director of programs in the life sciences, the mathematician Warren Weaver (1894–1978). The reasons for implementing such a program have been examined with a great deal of interest by historians. The development of biology appeared to be a priority. Problems that affected society, such as criminality and mental disorders, were considered to be biological problems whose solution would come from progress made in the realm of biology. Progress was thus not to be expected from taking hasty measures with little scientific basis, such as the projects

54. Hager, 1995; Morgan, 2013.

proposed by eugenicists, but from advances in our fundamental knowledge of the living world. But biology had somewhat lagged behind the other sciences, stuck as it still was in metaphysical debates. Its development had to be supported by the sciences that had experienced considerable transformations in the preceding years: physics and chemistry. Biology was to benefit both from the new models that those sciences had developed, and from the tools for the study of living things that they could help build.

The Rockefeller Foundation's analysis wasn't original: it was shared by many scientists. In 1927, well before the foundation's program was implemented, the physicist Jean Perrin (1870–1942) and Baron Edmond de Rothschild (1845–1934) created the Institut de biologie physico-chimique in Paris to promote research in biology by enabling that research to be supported by the work of chemists and physicists. To elevate biology beyond the views of Pasteur and Koch, who saw it primarily as being in the service of medicine; to transform the work of biologists so that it would no longer resemble that of stamp collectors; to search for the fundamental principles that would explain living phenomena and finally enable the remnants of metaphysics to be shelved—these were the goals shared by many biologists, but even more so by physicists. It was that revolution that was promoted, in various forms, by the Danish physicist Niels Bohr (1885–1962) and by the Austrian physicist Erwin Schrödinger (1887–1961) in his 1944 book *What Is Life?*

FROM GENES TO PROTEINS . . . AND BACK AGAIN

At the end of the 1930s, the new technology enabled the question of the chemical nature of genes to be broached. In 1937, William

Astbury (1898–1961) was the first to obtain an X-ray diffraction image of DNA strands, although the slide didn't reveal anything precise about their structure. DNA, known since the work of Miescher, was shown to be closely associated with chromosomes by the Swedish cytochemist Torbjörn Caspersson (1910–1997), working at the Karolinska Institute in Stockholm. But the significance of that association wasn't clear; many biologists thought that DNA might constitute a support for chromosomes, or a reserve of energy for their division—one could obtain a derivative of ATP, the energy currency of the cell, from DNA.

In 1944, Oswald Avery, at the Rockefeller Institute in New York, was the first to propose that genes were made of DNA. The research Avery was engaged in was unrelated to the chemical nature of genes.[55] Working on pneumococci, he confirmed the observations made in 1928 by the English doctor Frederick Griffith (1879–1941) that the bacteria could modify the nature of the sugar capsule that surrounds them by absorbing a "transforming principle" coming from other pneumococci. After 10 years of purifying and of physicochemical characterizations, Avery arrived at the conclusion that the transforming principle was a molecule of DNA.

Avery's results were received with perplexity. He himself hesitated to assert that the transforming principle was a gene. No one knew anything about the genetics of bacteria, and some microbiologists even thought that bacteria didn't have genes. As we have seen, the structure of DNA was still not well understood and the model of the tetranucleotide was incompatible with any specific role for the macromolecule.

However, Avery's results inspired further research. Using the new techniques of chromatography, Erwin Chargaff

55. Méthot, 2016.

(1905–2002) showed that the model of the tetranucleotide was false and that the percentages of the bases of DNA differ from species to species, although the quantities of adenine and thymine on the one hand, and of guanine and cytosine on the other, are always equal. André Boivin (1895–1949), working at the Pasteur Institute in Marnes-la-Coquette, near Paris, at the end of the 1940s showed that the quantity of DNA is the same in all cells of an organism except the cells of the germinal lineage, where there is only half, after meiotic reduction—a very strong argument in favor of a genetic role of DNA. In 1952, eight years after Avery's experiment, Martha Chase (1927–2003) and Alfred Hershey (1908–1997) confirmed the genetic role of DNA with a bacteriophage as their object of study. They used a method that radioactively marked proteins and DNA; this method was a "fallout" of the American nuclear program, and revealed that when a bacteriophage infects a bacterium and reproduces inside it, only its DNA penetrates the bacterium. Thus, it is DNA, not protein, that carries the genetic information enabling the reproduction of the bacteriophage.

The following year, Francis Crick (1916–2004) and James D. Watson (1928–) proposed a model for the structure of DNA.[56] Maurice Wilkins (1916–2004), working at King's College London, continued where Astbury had left off, and obtained higher quality images. It was Wilkins's work that convinced Watson and Crick of the importance of DNA and of the need to determine its structure. Wilkins was joined by Rosalind Franklin (1920–1958), who obtained images of even higher quality—images that, without her knowledge, enabled Crick and Watson to develop their model. Crick and Watson didn't perform any experiments themselves, but made use of the observations of others. How-

56. For Crick, see Olby (2009).

ever, their contribution was not unimportant. X-ray diffraction of DNA strands gives only indirect indications of the molecular structure. Watson and Crick were able to interpret those indications, combine them with other physicochemical information from the scientific literature, and by constructing a physical model—thereby imitating the strategy developed by Pauling—determine the structure of the macromolecule.

The proposed double helix structure remained a model, and it would take almost 30 years for other techniques to confirm that DNA did indeed adopt that structure inside cells. In the meantime, other forms, such as Z-DNA (z for zigzag), had their hour of glory.[57]

All the proposed structures nevertheless shared characteristics: the complementarity of the bases present on the two chains of DNA, adenine always facing a thymine, and cytosine a guanine. This characteristic explained Chargaff's observations, and suggested a simple model of DNA replication: the two chains separate, and each chain interacts with the complementary bases present in the cell to form a daughter molecule identical to the original molecule. This model was confirmed in 1957 by Matthew S. Meselson (1930–) and Franklin W. Stahl (1929–) using experiments of ultracentrifugation.[58]

The discovery of the structure of DNA was picked up widely by the media, inspiring designers and architects. Without doubt, the regularities and symmetries of the structure endowed it with a certain beauty. Above all, it represented a magnificent illustration of the way in which a structure can illuminate a function—in this case, genetic function. It explained how a molecule could make a perfect copy of itself, but also the

57. Morange, 2007.
58. Holmes, 2001.

ability of genes to determine the structure of proteins, as we will see below.

Determining the structure of proteins required much more time.[59] Since Fischer, it had been accepted that proteins were the result of the linear chaining of amino acids linked by peptide bonds; and that this long chain folded on itself to acquire its shape, or "conformation." Most biochemists agreed, however, that there must be simple rules determining the order of the amino acids and the folding process.

While working on fibrous proteins, such as keratin, Astbury had shown that their structure was repetitive, and he proposed a model. In 1951, Pauling and Robert Corey (1897–1971) corrected Astbury's results and proposed two models of recurring arrangements of amino acid chains in proteins—the α-helix and the β-pleated sheet—models termed secondary structure (primary structure being the amino acid sequence), and demonstrated their occurrence in proteins.

A knowledge of secondary structures did not, however, reveal the conformation—tertiary structure—of proteins. It was necessary to make the X-ray diffraction spectra "speak," and to achieve that, certain problems needed to be resolved, such as the "phase problem." It was also necessary to perform a large number of calculations, made possible by the arrival of computers in laboratories. In 1957, John Kendrew (1917–1997) revealed the structure of myoglobin (a protein that stores oxygen in muscles) and, a few years later, Max Perutz (1914–2002) proposed the structure of hemoglobin.

The establishment of a relationship between genes and enzymes (proteins) began very early, in 1902, when the physician Archibald Garrod (1857–1936), with the help of William

59. Chadarevian, 2002.

Bateson,[60] showed that alkaptonuria—an anomaly more than a disease, in which the urine rapidly turns black—was due to a recessive mutation and probably related to the absence of an enzyme. Analogous examples correlating a gene mutation and a lack of metabolic reaction (and thus, probably, the absence of an enzyme to catalyze it) were obtained repeatedly. Muriel Wheldale (1880–1932), a student of Bateson, showed in 1916 that the synthesis of anthocyanins, plant pigments, is controlled by genes.[61] At the end of the 1930s, Ephrussi and Beadle arrived at the same conclusion regarding the pigments in fruit fly eyes.[62] In 1940, Beadle and Edward Tatum (1909–1975) went in the opposite direction, starting with known metabolic pathways and moving to genes, working with the fungus *Neurospora*. They showed that each step in metabolism is controlled by a different gene, and in 1941 they generalized their results by proposing the "one gene–one enzyme" hypothesis.

This hypothesis didn't specify the nature of the relationship. It was interpreted by some geneticists as a sign of identity, signifying that genes *are* enzymes. After Avery's results, and especially the revelation of the double helix structure of DNA, the problem was framed differently. In their second article on the structure of DNA, Crick and Watson suggested that the sequence of bases in DNA is the code that determines the sequence of amino acids in proteins. Indeed, the structure of DNA is almost completely independent of its base composition: it is a double helix of constant pitch and diameter. What distinguishes two structurally identical DNA molecules is the sequence of bases on the two chains—it is that sequence that

60. Harper, 2005.
61. Richmond, 2007.
62. Sapp, 1987.

must be recognized for DNA to be able to play its genetic role. The following year, the physicist George Gamow (1904–1968) proposed a stereochemical code for DNA molecules, showing how different amino acids could lodge in the cavities that form between the bases of DNA. Crick soon demonstrated that Gamow's code was incorrect, but the latter's attempt nevertheless pushed Crick to try to resolve the problem of the code theoretically.[63] Although that work of reflection didn't lead to a deciphering of the code—that is, to discovery of the rule to convert a sequence of bases into a sequence of amino acids (in particular because Frederick Sanger [1918–2013], thanks to the sequencing of insulin, showed that the sequence of amino acids didn't follow an obvious regular pattern)—it did lead Crick in 1957 to propose what he unfortunately called the "central dogma" of molecular biology,[64] many of whose predictions proved to be correct. Between DNA and proteins there is an intermediary, which he believed was probably RNA. The role of RNA had been suggested in 1937–1939 through the cytological studies of Torbjörn Caspersson, which showed a relationship between the rate of protein synthesis and the quantity of cytoplasmic RNA, a relationship that was later confirmed by the work of Jean Brachet (1909–1988) in Brussels. Crick recalled the hypothesis that the order of bases in DNA must determine the sequence of amino acids, and proposed that the folding of proteins was a spontaneous process, a direct result of the order of amino acids. The translation of the sequence of nucleotides of RNA into a sequence of amino acids in the protein was carried out by an adapter that paired with RNA and brought an amino acid with it. In 1961, collaborating with Syd-

63. Kay, 2000.

64. Crick, 1958.

ney Brenner (1927–2019), Crick determined experimentally the number of nucleotides necessary to code an amino acid. For that experiment, Crick and Brenner used a mutagenic chemical agent that enabled the insertion or deletion of a nucleotide (and thus of a base). They showed that the insertion or deletion of three base pairs was less disruptive to gene function than the insertion of one or two base pairs. They explained this a priori surprising observation with the hypothesis that three nucleotide bases code for one amino acid. The addition or deletion of three nucleotide bases thus adds or causes the disappearance of an amino acid, whereas an alteration by one or two bases disrupts the code and changes all the amino acids beyond the point of modification.

Deciphering the code, however, was made possible thanks to a completely different experimental approach: the development of in vitro systems of protein synthesis by Paul C. Zamecnik (1912–2009) at Harvard. This required the separation of the various cell components through ultracentrifugation.[65] The work of Albert Claude (1899–1983) and especially that of George Palade (1912–2008) showed the importance of ribosomes, small intracellular particles, in the synthesis of proteins. This work benefited from advances in electron microscopy, made possible, as with optical microscopy a century earlier, by the development of new methods for mounting and dying biological samples. It was thanks to this system of in vitro synthesis that Zamecnik's collaborator, Mahlon B. Hoagland (1921–2009), was able to characterize a small RNA required for protein synthesis (transfer RNA) whose properties were (approximately) those of the adapter described earlier by Crick. In 1961, Marshall W. Nirenberg (1927–2010) and Heinrich Matthaei

65. Bechtel, 2008.

(1929–), working at the National Institutes of Health in Bethesda, Maryland, added to the medium for protein synthesis—which contained, in particular, ribosomes and transfer RNA—a synthetic RNA consisting of a monotone sequence of uracils. It had been expected that an artificial RNA would be formless and thus inactive, but this RNA caused the synthesis of a protein formed of only the amino acid phenylalanine. Severo Ochoa (1905–1993) and Har Gobind Khorana (1922–2011) used other artificial RNAs to advance decryption of the code. Marshall Nirenberg developed a different, more direct method consisting of revealing the complexes formed between ribosomes, trinucleotides, and transfer RNAs charged with radioactively labeled amino acids to complete the earlier results. The code was finally deciphered in 1965, after five years of effort and competition among laboratories, followed almost day to day by the American media. The fact that 20 amino acids are coded by some 60 codons corresponds to what has been called codon degeneracy. Crick proposed the wobble hypothesis of pairing to account for it: the third base of a codon can form nonstandard pairings with the corresponding base of the transfer RNA, allowing a single transfer RNA to interact with several different codons—the consequence is that the same amino acid can correspond to several codons.

Microbiology provided the last piece for the edifice of "classic" molecular biology, through description of the mechanisms that control the activity of genes (their expression).[66]

Going back to Félix d'Hérelle's ideas, in 1936 Max Delbrück (1906–1981), a German physicist who emigrated to the United States, chose the bacteriophage as the best biological system to reveal the secrets of living things—it was a simple organism

66. Cairns et al., 1992.

capable of reproducing in 20 minutes. Delbrück's earlier work in biology (in 1935) had consisted of inferring the properties of a gene from an analysis of the mutagenic effects of radiation. This was an experimental strategy that had been widely used by physicists since the beginning of the twentieth century: the effects of bombarding a target (atom, nucleus) with different kinds of particles can reveal a great deal about the structure of the target. This study was carried out in Berlin in collaboration with the geneticist Nikolay Timofeeff-Ressovsky (1900–1981) and the physicist Karl Zimmer (1911–1988). Erwin Schrödinger discussed it at length in his book *What Is Life?*

In 1942, Delbrück and Salvador Luria (1912–1991), cofounders of the phage group,[67] demonstrated that bacterial resistance to bacteriophages was not the result of successive adaptations to infection, but came from the selection of rare, spontaneous, preexisting variations. Their experiment banished neo-Lamarckism from bacteriology, where it had a lot of support (for example, among the Pasteurians, who used it to explain the attenuation process used in producing vaccines), and sealed a strong alliance between molecular biology and the synthetic theory of evolution (see next chapter).

In 1946 at Yale University, Joshua Lederberg (1925–2008) and Edward Tatum demonstrated the existence of genetic exchange between bacteria. The mechanism of this bacterial "conjugation" was decoded by Elie Wollman (1917–2008) and François Jacob (1920–2013) at the Pasteur Institute in Paris between 1954 and 1957.

The main focus of these two researchers was the study of lysogeny, a strange phenomenon in which some bacteriophages remain "silent" inside a bacterium, then suddenly multiply and

67. Brock, 1990.

provoke the lysis of their host. Revealed shortly after the discovery of bacteriophages, the phenomenon was so "bizarre" that many microbiologists denied its existence and attributed the bacterial lysis to accidental contamination. In 1949, André Lwoff (1902–1994), director of the lab where Jacob and Wollman would work, showed that the lysis of lysogenic bacteria could be induced by radiation, or by the addition of various chemical agents. He thus offered a means to experimentally control the phenomenon. Using the tool of bacterial conjugation, after a few years Jacob and Wollman arrived at the conclusion that a bacteriophage was carried by the bacterial chromosome in an inactive form called a prophage. They showed that the maintenance of that inactive state was due to the genes of the bacteriophage and to the presence of an inhibitor within the bacterium.

In 1957, Jacob began a collaboration with Jacques Monod. Since the early 1940s, in the same microbial physiology lab, Monod had been studying a phenomenon that had been revealed by Pasteur and Duclaux at the end of the nineteenth century—enzymatic adaptation, where a bacterium (or a yeast) placed in the presence of a sugar such as lactose begins to manufacture the enzymes necessary to break down that sugar.[68] The collaboration between Jacob and Monod, initially a technical one—Jacob providing new genetic tools enabling the characterization of the different mutations disturbing the metabolism of lactose that Monod had isolated—became a close scientific collaboration when the two scientists realized that the same genetic mechanisms could account for both enzymatic adaptation and lysogeny. In 1961, they proposed a model of regulation called the operon model. Two different types of genes exist in

68. On Duclaux, see Morange (2006b).

the genome: structural genes that code for proteins having a function in the life of a cell, and regulatory genes that code for repressors whose only function is to bind to DNA upstream of the structural genes and to control their expression.

Our knowledge of the mechanisms that regulate gene expression has been considerably expanded since this early work. The operon model, however, immediately had an enormous impact. It specified the function of the so-called "messenger" RNA that serves as an intermediary between genes and proteins. It completed the central dogma—the description of relationships between DNA, RNA, and proteins—by showing how proteins regulated the expression of genes. It also opened the door to a molecular description of embryonic development. In 1935, in his work *Embryology and Genetics*, Morgan had asserted that the major problem for embryology was to understand the mechanisms that differentially regulate the activity of genes during embryonic development and the concomitant differentiation of cells. The operon model wasn't the solution, but it enabled scientists to imagine what that solution might be. It convinced many molecular biologists to abandon the study of bacteria and to turn to that of higher organisms and their development.

In the 1960s, when molecular biology was asserting itself, often at the expense of older biological disciplines that began to see both their intellectual and financial resources shrinking, its most obvious characteristic was its use of concepts from informatics to explain its models. The diagrams showing patterns of regulation of gene expression resembled the diagrams of cybernetic systems in vogue at the time. This characteristic of molecular biology was acquired gradually after World War II with the borrowing of cybernetic concepts that, with the rise of computing, had become an integral part of the culture of the time. The informatics view fostered the concept of a genetic code, and its discovery and decryption, in turn, supported that view. These cybernetic concepts formed the mold in which the new discipline was shaped, and which gave it its appeal.

THE RISE OF GENETIC ENGINEERING

In the 1960s, molecular biology had a wealth of models, but was incapable of validating them experimentally as it lacked the appropriate techniques, particularly for the study of complex organisms. The only technique available at the time was the hybridization of nucleic acid molecules developed by Sol Spiegelman (1914–1983) at the University of Illinois: this technique with great difficulty provided complex results on the organization of genomes and the nature of transcripts that were often difficult to interpret.

The landscape changed completely in the 1970s with the development of tools for genetic engineering. Restriction enzymes, which enable a DNA molecule to be cut at precise locations, were purified. "Vectors" (small DNA molecules enabling genes to be cloned and reproduced inside bacteria) were created; techniques for sequencing and for the directed mutagenesis of DNA were developed; the first experiments in transgenics, in both plants and animals, were carried out at the end of the decade. The development of each of these techniques was important, but it was the progressive construction of an entire network that endowed each with its full potential. The techniques of molecular biology were then rapidly adopted by all biological disciplines.

In the 1980s, the possibility of making bacteria or yeasts produce human proteins for use as medication—insulin, interferon, growth hormone, blood clotting factors—at the moment when the AIDS epidemic revealed the potential dangers of extracting active substances from the blood (or organs) of other human beings, caused an explosion of new biotechnology companies.[69]

These developments, however, also provoked opposition and fear, especially since one of the first experiments to use the new tools, carried out by Paul Berg (1926–) at Stanford University, consisted of having a cancerous virus multiply inside a bacterium. A moratorium was called, and the Asilomar Conference in 1975 led to the adoption of a number of containment rules, although these would be gradually relaxed, even abandoned, as experiments continued to multiply.

Historical Overview

The Complex Dance of Disciplines

During the first half of the twentieth century, as we have witnessed, there was a restructuring of biological disciplines. New disciplines, such as genetics and biochemistry, were formed. Others, such as physiology and cellular biology, became less visible, despite the important work of neurophysiologists, and Barbara McClintock's and Cyril Darlington's observations on chromosomes. Even recently formed disciplines had difficulties: population genetics, which we will discuss in the next chapter, was rather quickly formed into a discipline that was almost independent of genetics. Molecular biology expanded beyond and in opposition to biochemistry. In this complex landscape, two disciplines deserve particular attention: genetics and molecular biology.

69. Hughes, 2011; Rasmussen, 2014.

Genetics developed all the more quickly as it was isolated from the other biological disciplines, and it defined a research program that was all its own. This was the case in the United States, where the analysis of the different allelic forms of genes and their positioning on chromosomes left the question of their physiological role in second place. Genetics gradually assumed a dominant position among the biological sciences for two distinct, but connected, reasons. The first is that to many observers it appeared to be the "physics" of biology, the only branch of biology that was experimental, had established laws, and analyzed its results with the use of mathematical tools. The second is that the very object of genetics, the gene, turned out to be for biology what the atom is for physics.

For Muller, the gene was at the origin of life: in his opinion, the appearance of life could be identified with the formation of genes. Genes remained unaltered by their passage through organisms over generations. They were of a remarkable stability: to illustrate that stability Schrödinger chose the example of the genic form responsible for the Hapsburg lip, recognizable in portraits of the kings and princes of that family, which dominated a large part of Europe for several centuries. In a lecture he gave in 1945, Muller used several metaphors to describe the essential—but parsimonious—action of genes in an organism: they are the orchestra conductors, points of reference, beacons that guide the other cellular components as radio waves guide planes. The both structural and functional characteristics attributed to genes made it difficult to identify them as objects. Whereas the establishment of gene maps and the confirmation of their "reality" through cytological studies contributed to a process of objectifying the gene—in the literal sense of transforming a gene into an object—Morgan could still claim, in his Nobel speech of 1934, that genetics would remain unchanged if the positioning of genes on chromosomes proved to be an illusion. This disincarnate side of genetics would make the emergence of molecular biology more difficult: the identification of the gene with the DNA molecule, whose structure, in addition, was considered chemically monotonous, would take several years. In his autobiography, Jacob said that when he first told Monod of his hypothesis that the repressor interacted directly with genes, Monod—convinced since the 1930s of the importance of genetics—was adamantly opposed to the apparent undermining of the "extraordinary" character of the gene.[70] Belief in the abstract nature of the gene, which was sustained by its identification with a form of information, would also reinforce the belief in a strong genetic determinism.

The status of molecular biology is problematic. It is not, and never has been—except, perhaps, very fleetingly in the 1960s—a discipline in the institutional sense of the term. There are very few university departments with that name, very few scholarly journals, and there are no corresponding scientific societies. In terms of research programs, molecular biology is scarcely distinguishable from biochemistry, which well before molecular biology had undertaken the characterization of biological macromolecules. Microbiology became molecular by way of a quiet, imperceptible transformation, quite unlike the "revolution" sometimes used to describe the emergence of molecular biology. However, to reduce molecular biology, as many of its detractors do, to a set of techniques—those of genetic engineering—isn't satisfying either, if only because molecular biology was perceived as a new approach to the study of life well before those techniques were available. The American historian Lily Kay has described molecular biology most accurately as a

70. Jacob, 1988.

new vision of life.[71] It is the description of a level of organization of living matter—that of macromolecules—that was unknown before the discipline appeared. But this is not just any level, because it appears to be the most apt (or one of the most apt) to reconcile the structural and functional approaches—the study of the structure of macromolecules, of DNA and proteins, reveals their function.

Molecular biology couldn't have emerged without the development of a new branch of chemistry that assigned major importance to weak bonds. Notions such as conformation (and transconformation) were born out of that new chemistry. It enabled at least the partial reconciliation of the physical and chemical conceptions of life that had clashed with each other since antiquity. The success of the expression "molecular machines" to describe proteins or groups of proteins is proof of this.

The Identity of Objects Studied and the Tools for Studying Them

It is well known that the results of a scientific discovery quickly become the tools for new discoveries. As soon as the laser had been developed, it was used widely in physics experiments, because it enabled the production of an intense beam of monochromatic light. The molecules synthesized by chemists are often the starting point for new syntheses. The various mutations isolated in fruit flies quickly became tools for isolating other mutations, or for studying specific phenomena. There is nothing surprising in this: scientific knowledge is the result of interactions between human beings and their environment; the

71. Kay, 1993.

tools of this interaction are continually modified as a function of the results of earlier experiments.

Biochemistry and molecular biology do not escape this rule. Let's look at a few examples in the history of their development. When Sumner crystallized urease, he benefited from the existence of partially purified preparations of the enzyme that were already in use for assaying urea. When Beadle and Tatum used *Neurospora* to isolate and characterize mutations affecting certain steps of metabolism, the mutant strains they had developed—dependent on the presence of particular molecules, like vitamins, for their growth—they subsequently used to assay those molecules. To characterize the nature of the transforming principle, Avery and his colleagues used different physicochemical methods, but they also used partially purified preparations of enzymes specifically capable of breaking down proteins, RNA, or DNA.

The case of restriction enzymes is a very good example of modifying earlier discoveries for a new use. These enzymes are part of a complex system of modification/restriction, described in particular by the Swiss geneticist Werner Arber (1929–), which enables bacteria to break down exogenous DNA molecules from other microorganisms without altering their own DNA, which has been modified in a way that protects it.

As soon as it was demonstrated that some of these restriction enzymes cut DNA at specific sequences, they began to be used in gene cloning. Besides these enzymes, experiments in genetic engineering required an entire collection of proteins and enzymes whose normal functions were exploited, directly or indirectly, by the experimenter.

Without being entirely unique, this use of the outcomes of discoveries as tools, in biochemistry and in molecular biology, has two notable characteristics. First, it is the *object* of the discovery that is used, not a procedure or method developed out of the discovery. Second, it all happened on a large scale, and very quickly. As we have seen, these tools were often being used even before they had been fully characterized. Perhaps we should see here the legacy of that long tradition of using living beings themselves to prepare food, which goes back to the Neolithic and probably to an even earlier era. Biotechnology didn't wait for the term to be invented to exist.

Multiple Explanations—Contentious Explanations?

The emergence of the Darwinian theory of evolution would give rise to many situations where different types of explanation were used to account for the same phenomenon. Embryonic development was explained by Haeckel's theory of recapitulation, and by the proof of so-called "proximal" causes by the followers of the mechanics of development. Immunity was explained by the presence of antibodies and their interaction with their targets, as proposed by Paul Ehrlich, and also by the presence of cells whose function is to eliminate foreign elements introduced into the organism. The existence of "evolutionary tendencies" could be explained by laws such as those proposed by Cope, or by the permanent action of natural selection in an unchanged environment, as was very quickly asserted by Darwin and Huxley. In all of these cases, the opposing explanations have very different natures. It is nonetheless sometimes possible to bring them together. For example, in the case of evolutionary laws, one might justify their existence as a consequence of the action of natural selection in a specific environment. But the search for agreement is rarely successful. Two cases might be presented. One of the protagonists claims that the other's explanation explains nothing. This is what Paul Baumgarten asserted when he claimed that

Metchnikoff's explanation was worthless because Metchnikoff didn't describe the mechanisms by which phagocytes recognize their targets and eliminate them. Another example is when His asserted that the theory of recapitulation was not an explanation for embryonic development. For His, the only explanation was the description of mechanisms that act at each stage of embryogenesis. In that case, one of the protagonists rejected one type of explanation in favor of another type (mechanistic), but without, however, justifying his choice.

But another situation can arise in which the protagonists don't see the difference in the nature of the proposed explanations, and try simply to show that the other explanation doesn't agree with the facts. This is what many Darwinians did when they asserted that the so-called laws of evolution were worthless in general terms. It is also what American biologists who worked at the Woods Hole marine lab did. By determining the cellular lineage of multiple species of invertebrates, they showed that the theory of recapitulation didn't account for the observations they had made: some characteristics of development can be explained only by the adaptation of embryonic development to the conditions in which it occurs.[72] The mechanistic approach was stronger, not because it was superior in principle, but simply because the other hypothesis didn't account for all the phenomena observed.

For a biologist today, it seems incomprehensible that the theory of recapitulation could have been used as an explanation; at best, it might have pointed to the discovery of particular mechanisms of development. But that wasn't the case at the end of the nineteenth century, when the dominance of mechanistic explanations wasn't as strong as it is today. That an organism followed a certain path of embryonic development was a fact

72. Maienschein, 1978.

that didn't call for explanations: that was the case for Claude Bernard, who was indifferent to the existence of a process for the evolution of organisms, and was convinced that this path of development was simply transmitted from generation to generation; as it was for Haeckel, whose entire work was centered around the idea of evolution. What has to be explained at any given moment is the result of a particular episteme, but also depends on the type of scientific approach that has been chosen. A paleontologist, whose work consists of comparing different fossils, will favor a search for laws enabling him or her to connect the different fossils under study.

Today, one finds analogous cases of opposition among different types of explanations: The domain of phyllotaxy, which studies the arrangement of leaves on stems, continues to experience clashes, as in the nineteenth century, between different explicative models, physical and molecular. Similarly, theories of a role of mechanistic forces in the first stages of embryonic development come up against more "molecular" explanations.

Embryonic Induction, Hormones, and Genes: Another Model for the Action of Genes

The change that occurred between the first experiments of Beadle and Ephrussi on determination of eye color in fruit flies and the implementation of the *Neurospora* system to characterize mutations affecting stages of metabolism is often misunderstood, and, for that reason, its importance understated.

That change corresponds to the abandonment of an explicative model of development that was at the origin of work on the fruit fly, as well as of Kühn's work on *Ephestia*.[73] Genes enabled

73. Rheinberger, 2000b.

the synthesis of the hormones, which had an inductive role in the course of embryonic development. This model thus took into account the recently demonstrated importance of hormones, the role of "controllers" associated with genes, and the results of experiments in embryonic induction carried out by Spemann.

But the results did not support that model. The substances that were isolated, in particular by Butenandt, were not hormones, but intermediaries of metabolic pathways.

The prominence this model achieved, and its rapid abandonment, provide two lessons. The first is that scientific models are often the result of temporal convergences: the simultaneity of work showing the importance of hormones and the major role of induction phenomena during embryonic development, as well as the concomitant rapid rise of genetics, called for such a model.

The second lesson is that even if a model has been imposed by a rapid succession of observations, it is nonetheless not guaranteed to "hold" if experiments don't provide the expected results. Beadle, and most other molecular biologists, quickly changed direction, leaning toward another conception of the function of the gene, and leaving observations on the action of hormones in the differentiation of tissues and phenomena of embryonic induction—which were, however, well substantiated—temporarily unexplained.

Contemporary Relevance

The Recurrent Enigma of Phenomena of Regeneration

Abraham Trembley's discovery of regeneration in the hydra led to many experiments, and the discovery assumed an important, but ambiguous, place in the debate between epigenesis and preformation.

At the beginning of the twentieth century, using the new approach of experimental embryology, Morgan attempted to shed light on the subject, but without much success. In the 1960s, the regeneration of planarian worms was used to demonstrate the existence and determine the nature of "memory molecules," which were believed to store acquired behaviors in a coded form (see chapter 10). The experiments carried out at the time led to more confusion than results. Today, in the time of regenerative medicine and stem cells, conditions appear to have come together to be able to clarify this phenomenon, which up to now has remained a mystery. But the large amount of work that has been carried out recently seems to have divided this phenomenon into a set of different mechanisms.

Regeneration is not a unique case: suspended life, the ability some organisms have to slow their metabolism and sometimes even to dry out, like rotifers, waiting for more favorable conditions,[74] appeared to Spallanzani, and then, a century later, to Claude Bernard, as a window opening on to the most fundamental mechanisms of life. Though new methods for "suspending" life were developed in the twentieth century (through freezing in liquid nitrogen) and are widely used in medically assisted procreation, very few scientists today choose to make that phenomenon their main focus of research. From all the studies that have been carried out, recipes—not principles—emerged.

The death of cells through apoptosis is another example of such false "good" paths of research. Observed at the end of the nineteenth century, apoptosis was described with precision in the 1970s. It doesn't involve the accidental death that might strike a cell, for example, following oxygen deprivation, but a so-called "programmed" death, useful to the organism—like,

74. Tirard, 2010.

for example, the death of certain cells that enables the "sculpting" of an embryo. At the same time that apoptosis was described, genetic mechanisms enabling it to be carried out were discovered and decoded in the nematode *Caenorhabditis elegans*, which, under the direction of Sydney Brenner, became a model for developmental biologists in the 1970s.

The model of apoptosis remained important, but it quickly lost its original simplicity. The distinction between programmed and accidental cell death proved more porous than previously imagined. Furthermore, it was demonstrated that programmed cell death corresponded to a set of different mechanisms.

Regeneration, suspended life, and cell death didn't prove to be "solvable problems," as the immunologist Peter B. Medawar (1915–1987) had described them, or more precisely, problems whose solutions would lead to developments in all the biological disciplines. Perhaps that is because these three phenomena—which attracted the attention of biologists because they seemed capable of resolving questions of life (and death)—were questions that, with the whiff of philosophy they exude, do not belong to the realm of biology.

From Data Science to Networks

We have shown the role that concepts in informatics gradually assumed in the interpretation and presentation of results in molecular biology. This wasn't a transfer of concepts from one discipline to the other, but a gradual impregnation of molecular biology by concepts coming from the ambient culture.

One is hard-pressed not to make a connection with the growing role networks have assumed in the descriptions and explanations of contemporary biology. We no longer speak of pathways, but of metabolic networks. Signaling networks enable cells to respond to signals coming from their environment, and embryonic development is controlled by networks of developmental genes. The use of the term "network" is now common in evolutionary biology and ecology.

As in the case of concepts from informatics, it is not obvious that it was the efforts of certain specialists to apply network (or graph) theory to biological systems that led to their use. And attempts to attribute to biological networks precise characteristics that would explain their functional properties were received with little enthusiasm, and sometimes violently criticized. Perhaps we should simply see in this use of the network metaphor by contemporary biologists the consequences of the influence that social networks and the Internet have in contemporary culture.

Metchnikoff, the Inventor of Exaptation?

Metchnikoff interpreted the antimicrobial action of phagocytes as a conversion of their primary evolutionary function, which was to nourish organisms that didn't have a digestive system. This description is a superb example of exaptation, as Stephen Jay Gould explained the term: the acquisition by a biological structure of a new function through the action of natural selection, independent of the selection pressure that originally engendered the structure. One might question the reality of the evolutionary process that Metchnikoff described; but the description he gave of it is clearly one of exaptation—100 years before Gould. The history of science is interesting not just because it reveals the existence of pioneers, but more often because it sheds doubt on overly strong assertions of novelty.

The Explanation of Diseases: A Plus or a Minus?

Within a few decades, two opposing models were proposed to account for disease: infectious diseases were explained by the introduction of a pathogenic microorganism; deficiency diseases were the result of the absence of an essential element, a vitamin, in food.

We have seen how difficult it was to go from one model to the other, to abandon the search for pathogenic microbes responsible for scurvy or beriberi. The desire to find those pathogens was so great that many results in that area were published.

These two explicative possibilities belong to the toolbox of the contemporary biologist. Both types of explanation coexist in the case of cancer. The multiple genetic changes necessary for the formation of a cancer cell cause losses of function (due to the inactivation of genes that suppress tumors), but also gains in function, corresponding to amplifications of genes or to mutations increasing the activity of the proteins encoded by those genes.

In the case of aging, there are also explanations by loss of function (the largest number) coexisting with explanations by gain in function. Aging could be due to shortening of the extremities of chromosomes, called telomeres, to the at least partial loss of mechanisms of adaptation to variations in the environment, or to the disappearance of stem cells that enable the renewal of tissues. But, as we have seen, some researchers see aging as the result of an accumulation of protein aggregates, toxic for the cells in which they form. The two concepts are far from being incompatible. Protein aggregates can impair cells and alter some of their functions. Nevertheless, the origin of the dysfunction—that is, its cause—should ideally be attributed to one or the other of the two types of explanatory phenomenon.

What Are the Colloids of Today?

The notion of macromolecule gradually gained ground, to the detriment of the colloidal theory of life. We must not, however, deduce from this that colloids don't exist, and that the many Nobel Prizes awarded for work on colloids were for naught. Today, we still find articles on colloids published in excellent scientific journals, and the theory of colloids is frequently used in some areas of chemistry. What has completely disappeared, however, is the impetus to make colloidal theory the explanation for phenomena of life. Just like the term "protoplasm" before it, the term "colloid" seemed in itself to represent an explanation, and to dispel the need to investigate the "how" of described phenomena any further. This ability of some notions or concepts to artificially fill an explicative void is rather common, at least when we look at them retrospectively. Which of the concepts that are used today will in 50 or 100 years prove to have played the same role? Terms like "system," "complexity," or "epigenetics"?

The End of the Dominant Position of Genetics

Genetics rapidly assumed a prominent place among the biological sciences, for the many reasons that we have discussed, and especially because of the very nature of the characteristics of the gene: purity, stability, indivisibility, essential but limited action, and the identification of its appearance with that of the first forms of life.

None of these characteristics has held up through all the work that has been carried out on genes throughout the twentieth century. Genes can be altered by other genic forms with which they have come into contact—these are called paramutations. The stability of genes is relative; it is the result of repair processes undertaken by cells, and is not due to the nature of a gene. Genes are not indivisible: intragenic recombination has been described since the 1950s. The action of genes is not necessarily limited: some genes are expressed in a permanent way. Genes as we know them, formed from DNA, were probably invented through evolution well after the first living beings appeared.

Given all of this, genes, and consequently genetics, should have been knocked off their pedestal. That has indeed happened in many fields of modern biology, in which genes are considered to be simply components of a more complex system. There are three things, however, that have held back, and continue to hold back, this movement of desacralizing genes. First, the gene, by its simple chemical nature, and since the invention of tools of genetic engineering, is the primary gateway to knowledge of living things. A comparison of DNA sequences is the best way to establish phylogenies, and the introduction or inactivation of a gene constitutes the simplest means to test the function of the macromolecular components of an organism, whatever they may be. The second reason is the role of genes in pathological processes. It was at the time (in the 1980s) when the gene was losing its preeminence that the isolation of genes linked to diseases became possible thanks to the tools of genetic engineering, and genes began to take on an increasingly important role in research. The third reason, a consequence of the second, is that genes continue to have a high profile in the media, as well as in education. The terms "gene" and "DNA" are increasingly used to describe what is fundamental and characteristic of an institution or a company—the "corporate DNA." The use of DNA by police scientists to identify individuals probably also has much to do with this. With their growing use, these metaphors support an inexact conception of the "power" of genes. For all these reasons, genetics has not yet fallen off its pedestal.

The Asilomar Conference: A Model?

The Asilomar Conference and the moratorium that preceded it are often cited as examples of responsible oversight by the scientific community of the risks linked to the implementation of new technology. In the face of the development of synthetic biology—that is, the manufacture of artificial living organisms—some people are demanding a moratorium and a new Asilomar.

That model has limits, however. The first is that the measures adopted were very quickly abandoned, once the first experiments were carried out. The second is probably more important: Does the decision of whether or not to develop a new technology belong exclusively to the scientific community? Asilomar was a gathering of scientists, and didn't include any representatives of other segments of society.

9

The Twentieth Century (Part II)

THE THEORY OF EVOLUTION, ECOLOGY, ETHOLOGY

The Facts

Genetics and the Theory of Evolution (1900–1920)

The rediscovery of Mendel's laws, and the rapid development of genetics caused the balance to tip in favor of those who supported a discontinuous conception of evolution, as opposed to the gradualist concept of Darwinism, which had been picked up and modeled by biometricians. Indeed, in a continuous view of variation, any deviation seemed inevitably bound to regress toward the mean through blending. The mutation theory, published by de Vries in 1901–1903, was an immediate success.[1] The fact that its creator was one of those who had rediscovered Mendel brought the two concepts together.

However, de Vries had problems reconciling his view of mutation with the transmission of variation as described in genetic theory. Bateson didn't experience the same difficulties, and saw

1. Allen, 1969.

in the different allelic forms of genes sufficient elements of discontinuity for a theory of evolution.

There were many difficulties that could be overcome by the mutation theory, which explains its popularity among many biologists, including Morgan and Loeb. It allowed for a rapid evolution, which destroyed the objection raised by Lord Kelvin (1824–1907) concerning the age of the Earth—it wasn't as old as had been thought up to then. It also resolved the difficulty that the appearance of new species represented for any continuous variation theory. For de Vries, a single mutation was enough to create a new species. Since, in his view, mutations were produced in waves, the same variations appear simultaneously in males and females, eliminating the difficulty posed by the possible sterility of hybrids between mutants and nonmutated organisms. The mutation theory also accounted a contrario for the stability of the pure lineages studied by Johannsen.

Above all, mutation theory led to a surge of research on evolution; as the neo-Lamarckian theory had done before it (although the latter was rejected by the majority of biologists after Weismann showed that there was no heredity of acquired traits). It took the same experimental approach as was being used for embryonic development by Wilhelm Roux and embryologists. The theory could even account for orthogenesis: Why shouldn't mutations orient the evolution of organisms in the same direction?

The theory's apparent ability to resolve all difficulties was also the weakness of this saltationist approach to evolution. The mutation theory explains everything, so long as we ascribe to mutation characteristics that would not be confirmed by later experiments, such as those undertaken by Morgan on fruit flies.

The Rise of Population Genetics (1918–1932)

Population genetics enabled two difficulties to be overcome.[2] It was able to reconcile the observations of biometricians on regression and correlations with the discontinuous model of genetics. It demonstrated that the small allelic variations described by geneticists were sufficient to be sorted by natural selection. Results were obtained not through observation nor through experiment, but through simulation using mathematical models.

The first law of population genetics was proposed by the British mathematician Godfrey H. Hardy (1877–1947) in 1908. At Punnett's request, and to Hardy's great embarrassment, so mathematically simple was the demonstration, Hardy showed that the distribution of dominant and recessive forms of genes remained stable in a population in the absence of evolutionary influences, and in particular of any form of selection. This has become known as the Hardy–Weinberg law, because it was proposed in the same year by the German doctor Wilhelm Weinberg (1862–1937).

This first law showed the efficacy of mathematical modeling. Ronald Aylmer Fisher (1890–1962), John Burdon Sanderson Haldane (1892–1964), and Sewall Wright (1889–1988), in slightly over 10 years, established the foundations of the new field. Their results largely overlapped, although each provided an original contribution.

Ronald Fisher was an English mathematician, a specialist in statistics, trained in biometry. In an article published in 1918, he demonstrated that the observations of biometricians on the

2. Provine, 1971.

correlation of traits between parents and children, brothers and sisters, were completely compatible with a Mendelian model of heredity involving several genes for each trait.[3] The article is remarkable for its use of new statistical tools such as the calculation of variance, and for its scope: Fisher considered several cases—the existence of epistasis (nonadditivity of the contributions of different allelic variants), the presence of more than two allelic forms, and the existence of a link between different genes—and he showed that in none of those cases was his conclusion compromised.

In his 1930 work *The Genetical Theory of Natural Selection*, Fisher modeled phenomena of sexual selection, mimicry, and the evolution of dominance. He demonstrated that the probability of a mutation increasing "fitness"—that is, increasing the number of descendants—decreases proportionally to the magnitude of the effects of that mutation.

After studying at Cambridge, and with his discovery in biometry, Fisher became a confirmed eugenicist. He would remain one throughout his life, insisting, like Darlington, on the existence of differences among human races. Always ready to argue and to provoke, in the early 1950s he criticized the epidemiological research of Richard Doll (1912–2005), which showed a relationship between lung cancer and smoking. He called it a "simple correlation," such as the one that statistics seemed to show between the importation of apples into Great Britain and a rise in the number of divorces. And as we have seen, he also accused Mendel of "having arranged" his results.

Haldane's work wasn't limited to population genetics; his early work was in enzymology. In the various books and articles he wrote, he broached the question of the nature of the gene,

3. Fisher, 1918.

the relationships between genes and enzymes, and the origin of life. In 1924, at the same time as but independently of Aleksandr Oparin (1894–1980), he proposed the idea of a primordial soup at the origin of the first living beings.

In the realm of genetics, Haldane's early work consisted of giving a precise mathematical description of the expected result when two genes are linked. The aim of the series of nine articles that he published between 1924 and 1932 was to demonstrate that the theory of natural selection could quantitatively account for the facts of evolution. In particular, he showed that minor differences in "fitness" could nonetheless provoke a large and rapid displacement of the frequency of allelic forms. He provided a biochemical (enzymatic) interpretation of dominance, a subject that was widely debated among geneticists. He is also credited with converting the "arid" results of mathematical modeling into simple, demonstrative examples—such as the famous example of industrial melanism: moths changed color when the bark of trees in their environment became blackened from nearby coal burning, to retain their camouflage from predators.

Haldane's 1932 book, *The Causes of Evolution*, contains two distinct parts: a simple, illustrated presentation of many examples showing that "Darwinism is not dead," but on the contrary is capable of explaining facts of evolution observable in nature, and an appendix in which the mathematical theory of natural selection is presented.

A confirmed Marxist, Haldane left Great Britain in 1956 after the Suez campaign and spent the final years of his life in India, developing genetic research there.

The American geneticist Wright was a student of Castle; his early work dealt with the genetic control of coat color in mammals. Wright also provided a precise interpretation of phenomena of consanguinity. He was very quickly convinced of the importance of interactions among genes, and saw the recombination of allelic variants during sexual reproduction as playing a major role in evolution. Wright's most important contribution consisted of showing how four mechanisms at the origin of evolutionary transformations worked together: mutation, natural selection, genetic drift—a wholly random change in the frequency of an existing gene variant in a population—and gene migration, or gene flow. He used a specific model, the adaptive landscape, to explain the differential effects of the four mechanisms. In this model, organisms are laid out on a plane. The vertical axis represents the "fitness," which increases under the effects of natural selection. If that effect is too strong, it pushes organisms toward the closest adaptive peak, compromising their later adaptation should the conditions of their environment change. Unlike Fisher, Wright didn't believe natural selection was the only evolutionary force; he gave genetic drift an important, but complementary, role—it created new interactions among genes and thus new possibilities for the action of natural selection. Like Haldane, Wright sought to connect theoretical models to "reality," focusing in particular on the genetic control of enzymes.

At the École normale supérieure in Paris, Philippe L'Héritier (1906–1994) and Georges Teissier (1900–1972) attempted to verify the predictions of population genetics models, using the tools of experimental evolution to study fruit flies.[4] They used "population cages," which they developed for their research. Notably, they confirmed that unfavorable allelic forms could be maintained in a population if the heterozygous state confers a selective advantage.

4. Gayon and Burian, 2004.

The models of population geneticists—but probably even more so Haldane's and Wright's simple representations—convinced evolutionary scientists of the compatibility of Darwinism with genetics, and signaled the end of saltationism: Morgan's abandonment of it being the final signal.

Modern Evolutionary Synthesis (1937–1950)

Modern evolutionary synthesis is not the direct result of new observations. It is precisely what its name indicates—the pooling of observations made by geneticists in the laboratory; those made in the field in zoology, botany, and paleontology; and the mathematical models of population genetics.[5] Its ambition is lofty: to show that the evolutionary transformations of the entire living world can be explained though the mechanism of natural selection (along with other mechanisms such as genetic drift) acting on the allelic frequencies of genes. Evolution is gradual and macroevolution—the evolutionary changes whose traces can be observed by examining fossils—can be explained as the outcome of phenomena of microevolution as modeled by population geneticists.

Though it did not emerge from new discoveries, the modern synthesis is nonetheless supported by several lines of work. The description of fossils has progressed considerably since Darwin: their scarcity was the main reason for any weakness in his theory. Among the many discoveries, let's take that of American paleontologist Charles D. Walcott (1850–1927) in 1909 in the Burgess Shale in British Columbia, of Cambrian fauna more than 500 million years old—it would take several decades to

5. Cain, 2009.

obtain a precise interpretation of the fossils.[6] A description of the principal metabolic pathways provided by biochemists revealed their extraordinary evolutionary preservation, which, as we have seen, Kluyver stressed. The living world exhibits a profound unity, which is also found on the level of evolutionary mechanisms. In the first half of the twentieth century, the development of new methods in taxonomy—from the study of chromosomes, for example—gave new life to that field, and prepared it for an encounter with population genetics.

A look at the contributions of the principal actors in the modern synthesis is the best way to illustrate the convergence that this represents.

Julian Huxley (1887–1975) coined the term "modern synthesis" in his 1942 book *Evolution: The Modern Synthesis*. More than his observations on the importance of polymorphism and his work as a zoologist, it was because of his efforts to popularize science with the general public, and perhaps especially that he was the grandson of Thomas Huxley, that his work was considered to be the completion of Darwin's and his grandfather's project.

The American geneticist of Russian origin Theodosius Dobzhansky (1900–1975) in 1937 published a book titled *Genetics and the Origin of Species*.[7] Like Haldane, he sought to explain in simple terms the results of population genetics. He also helped connect observations in the laboratory—he worked with Morgan—with those made in the field. In this he was carrying on the work of the man who had been his professor in the Soviet Union, Sergei Chetverikov (1880–1959). At the same time as the British geneticist Edmund B. Ford (1901–1988),

6. Collins, 2009.

7. Dobzhansky, 1937.

Dobzhansky confirmed the importance of natural selection in the maintenance of genetic polymorphism and the advantage that heterozygosity can represent.

In 1942, Ernst Mayr (1904–2005) published his work *Systematics and the Origin of Species from the Viewpoint of a Zoologist*—a title that recalled both Darwin's book and that of Dobzhansky. In it he proposed that the principal mechanism of speciation was allopatric speciation—geographic isolation. Until his death at the beginning of the twenty-first century, Mayr would adamantly defend evolutionary synthesis, whose advent he would include as a landmark in the history of biology, and would argue for its central role in "biological thought."

George G. Simpson (1902–1984), professor of zoology at Columbia University in New York and curator of the Harvard Museum of Comparative Zoology, deconstructed what was one of the major obstacles to acceptance of the idea of gradual evolution, the evidence of paleontological data. In his 1944 book *Tempo and Mode in Evolution,* he showed using statistical methods that when the paleontological record is sufficiently complete, it doesn't reveal an evolution in leaps, guided by orthogenesis, but rather a bush-like evolution, the result of an accumulation of small-scale variations.[8] The absence of fossils of intermediary forms was explained by the fact that the evolving populations were small in size and that evolution was rapid. He showed that evolution could proceed through gradual changes within the same species, or through the appearance of new species.[9]

In 1950, the American botanist George Ledyard Stebbins (1906–2000) brought the plant world into the synthesis with

8. Simpson, 1944.

9. Laporte, 1994.

his work *Variation and Evolution in Plants*. Certain mechanisms, such as hybridization and the duplication of genomes (polyploidy), were of particular importance in the evolution of plants.

Scientific colloquia, and the newly created journal *Evolution*, were instrumental in the emergence of evolutionary synthesis. But even after meetings and the publication of papers, the synthesis was far from being completed, because its founders had very different opinions on its central questions.[10] Whereas Huxley and Stebbins were convinced that there is progress in evolution, Simpson was against that idea. There were also differences of opinion on the place of human beings in evolution. In the years after its introduction, the synthetic theory of evolution became "hardened." Natural selection, which for the founders of the new theory was one of the principal mechanisms of evolution, became the *only* evolutive mechanism, de facto excluding genetic drift, to which Wright attributed an important role. The distinction between the results of natural selection and the effects of genetic drift is not always obvious, but nor is it impossible to make, as shown by Maxime Lamotte in the 1950s.[11]

There would, however, be few true opponents of evolutionary synthesis, apart from a German geneticist who sought refuge in the United States, Richard Goldschmidt (1878–1958). He would never renounce saltationism. In his 1940 book *The Material Basis of Evolution*, Goldschmidt contrasted micro- and macromutation and micro- and macroevolution, and described macroevolution as the result of the appearance in one step of new forms, which he called "hopeful monsters."[12]

10. Delisle, 2009.

11. Millstein, 2008.

12. Goldschmidt, 1982.

Ecology

In the first half of the twentieth century, ecology was far from being a unified scientific field.[13] Specialists who studied lakes (limnologists) were not the same researchers as those who studied plant ecology, animal ecology, or oceanography. The heterogeneity of ecology thus in part originates in the diversity of what would be called ecosystems. The first half of the twentieth century saw the development of three different ecological approaches, which George Evelyn Hutchinson (1903–1991) would bring together in his Yale University laboratory in the 1940s and 1950s.

The first approach was within the tradition of biogeography, but gradually placed greater emphasis on the physical environment and the physiological properties of organisms. Plant associations, which would be called phytosociology, were studied in Uppsala, and by Charles Flahault (1852–1935) and his Swiss student Josias Braun-Blanquet (1884–1980) in Montpellier. So-called "faithful" species enabled the distinction of different types of plant associations. In 1895, the Danish scientist Eugenius Warming (1841–1924), showed how each member of a community exists in morphological, physiological, and anatomic harmony with the ecological conditions in which it lives. For Warming, the ecological conditions were as much, if not more, the physical conditions of the environment as they were interactions with other organisms.

Henry Chandler Cowles (1869–1939) was a student of Warming. In 1901 he described the system of the dunes located on the shores of Lake Michigan. This system had the advantage

13. Egerton, 1983; McIntosh, 1985; Drouin, 1991; Deléage, 1991; Hagen, 1992; Kingsland, 2005.

of "starting from scratch"—new dunes being formed constantly—and of thereby revealing the dynamic succession of several plant communities up to a state of dynamic equilibrium, which Cowles called the "climax." Cowles's description would later be supplemented by his study of animal and plant species together.

Frederic Clements (1874–1945) compared Cowles's succession of plant communities to the embryonic development of an organism.[14] Clements was convinced that, for ecology to be useful, it had to be able to reveal regularities in nature. A comparison between plant associations and an organism was for him the means to stress the extent to which the evolution of these associations is determined. The analogy was reinterpreted by the South African ecologist John Phillips (1899–1987) as "holistic"—the whole is more than the sum of its parts. The term "holism" was introduced in 1926 by General Jan Smuts, the prime minister of South Africa, and holistic and emergentist concepts were developed by several philosophers. But not everyone accepted this view. The American botanist Henry Gleason (1882–1975) attributed a major role to chance in the formation of plant communities. Arthur Tansley (1871–1955), professor of botany at Oxford, disagreed with both organismic and holistic views, and in 1935, after more than 30 years of dialogue and disagreements with Clements, responded by introducing the term "ecosystem."[15] He considered an ecosystem to be just one example of the artificially isolated systems that scientists study, a simple mental construction. Comparing the development of an ecosystem to that of an organism did not fully take into account the complexity of the former, which in-

14. Clements, 1916, 1936.
15. Van der Valk, 2014.

cluded inorganic elements, nor the possibility of several climaxes. Transformations of ecosystems also depend on human activity. In using the term "system," Tansley favored the quantitative study and modeling of ecological communities.

The English zoologist Charles Elton (1900–1991) is credited with promoting animal ecology, which had previously remained in the background owing to the great difficulty of estimating the sizes of populations. In particular, he helped to introduce the notion of energy in ecosystems, thanks to observations he carried out during an expedition to Spitsbergen with Julian Huxley in 1921. He showed that access to food is essential for the structuring of a biotic community. In his 1927 book *Animal Ecology*, he introduced several concepts that would assume increasing importance over the following years: the food chain and food cycle—with a precise description of the positions of organisms within them—and the notion of the niche, which he borrowed from the American zoologist Joseph Grinnell (1877–1939). He showed that the numbers and sizes of organisms at different levels of the food chain can be represented by a pyramid. He was not the first to point out that large predators were small in number and large in size but, like Newton and falling apples, he was the first to inquire into what had once been taken for granted.

The second approach began with the study of the dynamics of human populations, inaugurated by Malthus, and then turned to that of dynamics of animal populations. The Belgian mathematician Pierre-François Verhulst (1804–1849) modeled the growth curves, or logistic curves, of populations, which the American biologist Raymond Pearl (1879–1940) considered to be of great importance.

In 1897, the French entomologist Paul Marchal (1862–1942) had anticipated that this curve could not be applied in cases where there was a prey/predator relationship between two species, and had predicted the existence of oscillations. The question was investigated by the American physicist Alfred Lotka (1880–1949) in his 1925 book *Elements of Physical Biology*. The famous Italian mathematician Vito Volterra (1860–1940) would complete Lotka's work in 1926. His inspiration was the changes he observed in fish populations in the Adriatic Sea following the interruption of fishing during World War I.

As the title of his book indicates, Lotka's study was inscribed within a broader project: to bring together the scientific approach to the study of the living world and that of the inanimate world by using the same energetic models. In 1922, he described natural selection as a physical principle, as the fourth principle of thermodynamics, accounting for the evolution of biological systems.[16] The struggle for life is a struggle for energy, and the living organisms that survive are not simply the ones that know best how to capture and use that energy, but those that have succeeded in increasing the amount of energy that flows through them. Human beings, through their ever-increasing use of energy, only continue the work of evolution.

Following the work of Lotka and Volterra, the Russian biologist Georgy Gause (1910–1986) attempted to demonstrate experimentally, using various types of infusoria, the oscillations predicted by their models. He was a member of a very dynamic school of ecology that developed in the USSR in the 1920s.

The third foundational element of this new ecology was the creation of soil science, or pedology, by Vasily Dokuchaev (1846–1903) during his study of the black soil of southern Russia. He demonstrated that the nature of soil depends on the

16. Lotka, 1922a, b.

rocks, the physical factors of the environment (especially the climate), and the living organisms that transform the soil.

Vladimir Vernadsky (1863–1945) participated in one of Dokuchaev's studies. He subsequently worked in Paris with the chemist Le Chatelier, and later taught mineralogy. While in Paris, he had the opportunity to talk with the French Jesuit and paleontologist Pierre Teilhard de Chardin (1881–1955), who helped him to expand his thinking. Vernadsky wasn't the first to use the term "biosphere"—the Austrian geologist Eduard Suess (1831–1914) invented the term—but in 1926, in a work of the same name, he put it front and center. He saw the biosphere as a realm on the surface of the Earth occupied by transformers that convert cosmic rays into active terrestrial energy. Lamarck (and many others) had already demonstrated the role of organisms in transformations of rocks, but Vernadsky gave that transformation new dynamics. Like Lotka, he suggested that the more life progresses, the greater its power of chemical transformation; human beings just follow and amplify this trend.

Thus, the consequences of human activity weigh heavily on the biosphere. Svante Arrhenius had already discussed the "greenhouse effect" of carbon dioxide gas and the possible consequences of an increase in the consumption of coal. But it was Vladimir Kostitzin (1883–1963), a friend of Vernadsky, who would develop those ideas.

In the United States, the time was ripe for an increased interest in ecology: years of overcultivation and poor land management of the Great Plains of the central United States in the 1930s caused an ecological catastrophe, leading to the formation of clouds of dust as the topsoil was blown away by the wind—the "Dust Bowl." Hutchinson would be instrumental in developing the different currents of ecology: granting due importance to ecosystems, and quantifying and modeling them;

providing a formal definition of the ecological niche; and showing the disruption of the major ecological cycles caused by human activity.[17] We can distinguish two phases in his work. From 1930 to 1950, he studied the influence of physical and chemical conditions on communities of organisms, in particular those of lakes. He contributed directly to the use of radioactive markers in ecological work. After 1950, he attempted to describe the mechanisms at the origins of animal and plant diversity in ecosystems. He is considered the father of theoretical ecology, and had a talent for popularizing the new face of his discipline.

Hutchinson's students and collaborators played a major role in the emergence of this new ecology. The publication in 1942 of two articles by Raymond Lindeman (1915–1942) in the journal *Ecology* constituted a notable step forward. Chancey Juday (1871–1944) laid the groundwork in 1940 with his description of the energetic state of the ecosystem of Lake Mendota in Wisconsin. One of the articles Lindeman published two years later provided equations aiming to describe the dynamic flow of energy in such animal and plant communities.[18] Lindeman distinguished between producers of energy, consumers, and decomposers, and defined the productivity of each level of the system. He also studied the changes that can take place in an ecosystem. Despite its originality, the article was rejected by two reviewers who considered it too theoretical—not respecting the diversity of each ecosystem—and thought the theory wasn't sufficiently supported by observations in the field. It was only thanks to the personal intervention of Hutchinson that it was published after the death of its author. Today, Lindeman is considered the founder of the trophic-dynamic concept of ecology.

17. Slack, 2003, 2011.

18. Lindeman, 1942.

Eugene Odum (1913–2002) and his brother Howard (1924–2002), students of Hutchinson, introduced the most general formalization of Lindeman's model. In 1953, the brothers published *Fundamentals of Ecology*, which would be used by generations of students. Picking up Lotka's ideas, Howard Odum advanced the notion that the ecosystems best able to use energy flows are also the most capable of surviving and dominating. He applied this principle to human societies and their technology, converting ecology into a science whose domain of application extends well beyond the study of life.

The new energetic view of ecosystems didn't end their comparison to organisms, but it lost any metaphysical charge: it simply listed the resemblances between the dynamics of two complex systems.

The work that led to the emergence of the most important concept in ecology, that of the ecosystem, led the questions asked by ecologists to diverge from those asked by evolutionary scientists. On the one hand, ecologists favored the study of systems in equilibrium; on the other, they established a link with the physical characteristics of the environment and "functionalized" ecological systems as structures resulting from the flow of energy.

Ecologists didn't participate in the development of modern evolutionary synthesis of the 1930s. Was it necessary for its development for ecology to be set apart in that way, or was there a missed opportunity to bring ecology and evolution together?

Ethology

As with all scientific disciplines, there really isn't a founder of ethology. Aristotle was interested in animal behavior; Isidore Geoffroy Saint-Hilaire, son of Étienne, promoted the study of animal behavior in the Jardin d'acclimatation created by

Emperor Napoleon III; Darwin, in comparing the expression of emotions in animals and human beings, and showing the role of sexual selection, was working as an ethologist. And in the eighteenth century, Réaumur had provided a precise description of the social behavior of insects.

In the nineteenth century, the British amateur naturalist Douglas Spalding (1841–1877) began to study the relationship between learning and instinct. His work was discovered by Oskar Heinroth (1871–1945) in Berlin; Heinroth would be the teacher of Konrad Lorenz (1903–1989). At the same time, Charles Whitman, a cell biologist and founder of the Woods Hole marine laboratory, was devoting part of his work to the study of animal behavior. Julian Huxley might also be called an ethologist for his research on the behavior of birds.

Another early form of ethology may be found in Darwinian tradition and in Haeckel's forward-looking interpretation of it: that the first signs of psychological characteristics must be sought in single-celled organisms. An experimental approach of this reductionist bent, through microdissection, was undertaken in Berlin and then in Jena by Max Verworn (1863–1921).[19] In 1887, Alfred Binet (1857–1911), best known for his famous IQ test, discovered in paramecia a form of intelligence that he considered to go far beyond cell irritability. In articles published between 1897 and 1902, and in his 1906 book, the American zoologist Herbert Jennings (1868–1947) interpreted the behavior of paramecia as the result of a strategy of trial and error, and hypothesized that the organisms possessed a certain degree of consciousness.

William Wheeler (1865–1937) was an admirer of Réaumur. Influenced also by Whitman, he devoted a large part of his work

19. Morange, 2006a.

at the Harvard Bussey Institution to studying the social behavior of ants. In 1902 he coined the term "ethology" in an article in the American journal *Science*. For him, ethology was animal ecology—that is, the study of the way in which organisms adapt to their environment. He criticized contemporary attempts to extend plant ecology to the animal world, and thought it was ridiculous to talk about plant sociology. Whereas in plants adaptation to their environment is primarily physiological, in animals it occurs through the modification of their behavior. In his view, animal ecology, or ethology, should turn away from mechanistic and experimental biology and proceed through observation: he cited Goethe as a pioneer of this new approach.

This is the direction in which ethology developed, radically different from the behaviorist approach. For John Watson (1878–1958) and B. F. Skinner (1904–1990), who were the leaders of the behaviorist movement, behavior should be studied experimentally in the way Pavlov did: an organism was considered as a black box, with input being stimuli, and output the organism's responses.

In 1973, the Nobel committee acknowledged three men as the "founders" of ethology.[20] The first was the Austrian Karl von Frisch (1886–1982). He described the perceptive capabilities of bees and of demonstrated that they see "in color," but toward the ultraviolet end of the spectrum. Bees can orient themselves using the position of the sun, the polarization of light when the sun is hidden, and the Earth's magnetic field. Frisch is best known for his interpretation, published in the 1920s, of the dances of bees: the round dance, which signals the presence of a food source nearby, and the waggle dance, which indicates both the direction in which the food is located and its

20. Burkhardt, 2005.

distance.[21] Frisch didn't decipher the language of bees, however, just a single dialect, as different bee populations code information in different ways.

Frisch is sometimes excluded from histories of ethology because he didn't play a role in the institutionalization of the new discipline; however, his results contributed a great deal to its visibility. He revealed the extraordinary sophistication of the behaviors of organisms that up to then had been considered "simple." He was stripped of his position as professor in Munich by the Nazi regime, yet was convinced that it was necessary to adopt eugenic measures, and hoped that education would encourage people to accept them.

The Austrian zoologist, ethologist, and ornithologist Konrad Lorenz is the best known of the three Nobel laureates. Convinced that animals in captivity manifest an impoverished range of behavior, he turned his house into a sort of "open" menagerie, in which he studied the behavior of, in particular, the famous graylag goose. He believed that behaviors were established through an interaction between instinct and external triggering factors during a critical phase of development. He made the analogy between these triggering factors and the inductive effects that the Spemann school had described during embryogenesis. A link is formed between instinctive behavior and the external trigger, which Lorenz called the imprint. The experimenter could change the nature of the triggering factor, and thus also that of the imprint; instead of following their mother, the graylag geese followed Lorenz's boots, because they were the first things the geese had come in contact with after they were born.

Lorenz was a confirmed Darwinian. He considered behaviors as traits that had been selected, in the same way as physical

21. Frisch, 1953.

traits. They could be used for classification, and this comparative approach was particularly useful to the ethologist. But one could observe the same behavior—such as jealousy—in very different animals. Unlike most physical traits, a resemblance between the behaviors of different animals was, for Lorenz, more often the result of analogy than homology: it came from a convergence of adaptive processes.

The comparisons Lorenz made, in particular with human behavior, would provoke much criticism. On the one hand, he was criticized for demonstrating anthropomorphism—for projecting human behaviors onto animals. More seriously, his scientific convictions would lead him to strongly support eugenics projects, and to various political and social positions—for example, on the negative effects of marriage between individuals of different human groups—and to adhere to Nazism.[22] His later involvement in political ecology and the "green" movement, and his repeated assertion that he did not mistreat the animals on which he worked, do not easily cancel out the fact that he was accused, when a physician during the war, of having singled out individuals born of marriages between Germans and Poles, and deciding which of them had the right to reproduce—sending the others to concentration camps.

Niko (Nikolaas) Tinbergen (1907–1988), born in La Haye and imprisoned as a hostage during the war, later taught at Oxford. He gave ethology a much more respectable image. Unlike Lorenz, he worked in the field. Using the interpretative model proposed by Lorenz, he developed experiments that enabled him to substantiate it. He developed supranormal stimuli that induced innate behaviors in animals more effectively than natural stimuli: for example, plastic eggs that a bird would choose to sit on rather

22. Kalikow, 1983.

than its own eggs; or cardboard butterflies that males chose over females to mate with. The object of this work was to determine the characteristics of the stimuli that activate innate behavior.

Tinbergen also contributed to the establishment of ethology as a science by giving it a simple definition—the biology of behaviors—and by defining its goals. Ethology was to respond to four different types of questions:

(1) on the mechanisms of behavior: to define the stimuli that trigger it, and the mechanisms that make its achievement possible
(2) on the evolution of behavior with age: to define the first experiences required for a behavior to appear, and the stages and stimuli necessary for its development
(3) on the adaptive function of behavior
(4) on the formation of behavior during evolution.

Tinbergen was one of the first, after Ernst Mayr, to distinguish functional questions from evolutionary questions with precision, and among the latter, between questions concerning the adaptive value of variations and those dealing with the history of the establishment of those adaptations.

One of the major problems with Lorenz's and Tinbergen's interpretations of their results was a noncritical acceptance of the existence of innate behaviors, which they explained a bit too easily as inscribed within the genetic program, whose importance was simultaneously being demonstrated by molecular biologists. Lorenz made a distinction completely between acquired behaviors and innate behaviors, denying the possibility that in the course of evolution there could be a gradual passage from one type to the other.[23]

23. Brigandt, 2005.

Historical Overview

While looking at the scientific developments of the nineteenth century, we questioned the role that philosophical concepts, as well as the confrontation between science and religion, might have played. The twentieth century, much more than the nineteenth, was one of ideologies. At least three would interfere with the development of the sciences. And yet few historians have broached this topic with regard to the twentieth century, considering it more relevant to the nineteenth.

The Influence of Marxism

It is important to discuss the role that Lysenko's theory played in the USSR from the mid-1930s, and in all of Eastern Europe after World War II. Borrowing a number of agricultural practices developed at the beginning of the twentieth century by the Russian plant biologist Ivan Michurin (1855–1935), Trofim Lysenko used them to gradually develop a theory in opposition to genetics. In accordance with dialectical materialism, his theory held that one could transform organisms by modifying the environment, even to the point of being able to transform one species into another.[24] Lysenkoism was a diluted form of neo-Lamarckism. But that wasn't the basis of the success of the theory; success came because of Stalin's support.

The influence of Marxism, to which we will now turn, was one freed of any political pressure; it was located at the level of the conceptualization and explanation of biological phenomena.

Three realms deserve particular attention: population genetics, with the important role played by Haldane and the "Cambridge

24. Roll-Hansen, 2004.

Marxists"; Soviet ecology with Vernadsky; and the question of the origin of life with Oparin and, again, Haldane. Were these three scientific developments the result, in one way or another, of the attachment of their "birth fathers" to Marxism?

To be a Marxist could mean two rather different things: adhering to a political view of society and wanting science to be an agent of social and economic change, or placing science within the framework of dialectical materialist philosophy—that is, allowing the unity of opposites and going beyond negation as a principle of change either of the material world or of scientific knowledge.

After studying Haldane's work on population genetics during the years when Haldane had openly professed to being a communist, Simon Gouz concluded there was a certain Marxist influence in Haldane's scientific work.[25] Haldane rejected the habit of Soviet scientists of sprinkling their results with philosophical considerations that had nothing to do with the work carried out, and thought that a scientific theory, however in agreement it might be with dialectical materialism, should be judged solely on scientific criteria. But he broached certain questions concerning population genetics by adopting a perspective of constructive opposition—for example, between the rate of mutation and natural selection. This opposition wasn't resolved by denying the role of mutation or that of natural selection, but rather through a synthesis that required taking both agents into account. In this case, dialectical materialism seemed to have provided Haldane with a framework of thought that is well suited to the scientific questions he raised.

The same was true for the question of the origin of life. Dialectical materialism, at least superficially, enabled one to resolve an apparently intractable opposition between life and nonlife, and to propose a path to go from the latter to the former. Again,

25. Gouz, 2006.

dialectical materialism offered a useful framework for thinking, although here a rather feeble one.

Adhering to Marxism also prompted one to seek, or at least claim to seek, the usefulness of the scientific knowledge acquired. Like the activity of Vernadsky and his colleagues, who promoted a broad view of ecology that included the transformations humans cause in nature—which would lead Vernadsky to encourage the development of nuclear energy—Haldane's positive involvement with eugenics went in that direction.

The Rise of Holism and Emergentism

Marxism wasn't the only ideology biologists adhered to in the first half of the twentieth century. The "holists" were opposed to the dominant reductionist approach in biology and considered that the study of the "whole" was essential. Holistic ideas existed before the invention of the term "holism": those who didn't accept Broca's results, and rejected the hypothesis that functions were localized in particular zones of the cerebral cortex, were holists. Emergentism, described below, provided them philosophical justification by asserting that some characteristics of systems emerge from the integrated functioning of their parts. Organicism was a form of holism that used the organism as a model of those complex systems whose functioning could not be explained in terms of a description of their parts.

Driesch was, at least in part, at the origin of these debates. In asserting that embryonic development is oriented by a nonmaterial force, which he called entelechy, he exerted a not insignificant influence on many of the embryologists, such as Spemann and Weiss, who adopted an organicist point of view and believed that the study of embryonic development could not be reduced to the study of the individual components of the organism. The notion of a morphogenetic field is the perfect example of this nonreductionist view.

Paradoxically, emergentism was also born out of a reaction to Driesch's spiritualist interpretation. For the philosopher Samuel Alexander (1859–1938) it was a matter of including emergent phenomena in a materialist view of the world and, unlike Driesch, of naturalizing them. Philosophical emergentism had its heyday at the beginning of the 1930s. But one of the examples chosen by the emergentists was turned against them: the idea that the emerging properties of a molecule of water could not be reducible to those of its components, hydrogen and oxygen. Chemists quickly showed that these properties were seen as emergent only because of the insufficiency of the theories available to them. With quantum theory, these formerly emergent properties could be predicted, at least in part, from a knowledge of the elemental components.

Despite these philosophical debates, holism (and organicism), as we have seen, maintained their place in ecology until the 1940s, and even beyond. After World War II, holism regained a certain vigor in the 1960s with the discussions on artificial life and models of the living world proposed by Tibor Gánti (1933–2009), Robert Rosen (1934–1998), Francisco Varela (1946–2001), and Humberto Maturana (1928–). The notion of emergence in biology reared its head again at the end of the twentieth century with the development of systems biology. However, the difficulty in attributing a precise meaning to the term "emergence" is still with us!

The Energetics View of Life

There is a third ideology, less structured and apparently more "scientific," but just as influential for biologists, which consists of reducing an explanation of life to that of an energetic phenomenon. The ideology has its origin in the development of thermodynamics in the nineteenth century, but also in the monist thinking of Haeckel and his disciples. It doesn't consist in reconciling the characteristics of organisms with the principles of thermodynamics, but of making life itself a branch of thermodynamics. The physicochemist Wilhelm Ostwald (1853–1932), president of the monist alliance from 1911, was one of the proponents of this global energetics view. In his 1930 book, Fisher linked the action of natural selection to the second principle of thermodynamics. Lotka was the scientist who gave that connection its most precise form, making natural selection the fourth principle of thermodynamics and explaining the evolution of the entire living world by a single rule: an increase in the flow of energy through organisms. With Lindeman and Hutchinson, this concept was extended to ecology, reducing the description of an ecosystem to one of the flow of energy through the organisms that compose it.

The second principle of thermodynamics also inspired many biologists. Thus, the disappearance of living species, as well as the loss of biochemical functions during evolution, which André Lwoff described in his 1944 book *L'évolution physiologique: Etude des pertes de fonctions chez les microorganismes* (Physiological evolution: The study of loss of function in microorganisms),[26] were seen as a direct consequence of the second principle.

The risk of these various energetics approaches was of reducing the complexity of organisms and their evolution to a single dimension. The evolution of species could not be reduced to a question of metabolism, and even less to one of a loss of biochemical functions. At the beginning of the twentieth century, the biochemist Hopkins had sounded the alarm by showing how reducing a nutritional contribution to a contribution of

26. Lwoff, 1944.

energy (calories) had delayed the discovery of essential amino acids and vitamins. But 30 years later, that wouldn't prevent ecologists from enthusiastically welcoming the energetics view of ecosystems. It would take several decades to be able to explain how, despite such a simple determinism, an ecosystem could be so rich in living species. Hutchinson's scientific career reflects well these two successive phases of ecology.

Reducing living phenomena to energetic processes allowed easy circulation between biological models and those used in economics. Mathematicians like Lotka, participating in the development of both types of models, were instrumental in the sharing of these models. The abundance of economic terms used in evolutionary biology and in ecology is, in part, a consequence of this, although there are other contributing factors, such as the development of game theory after World War II.

The three ideologies that influenced biological models were by no means in opposition. It is difficult to say whether the ecologists assembled around Vernadsky were Marxists, holists, or energeticists—probably all three at the same time. None of those ideologies had the strength to give birth to new paths of research, but they could provide them an accommodating framework for thinking.

The Question of Life

The first half of the twentieth century was one of the moments in the history of biology when the question of life was raised most directly, before the question disappeared along with the rise of molecular biology. For the pioneers of molecular biology, who came out of disciplines other than biology, unveiling the secret of life was essential because the absence of clear answers enabled the survival of metaphysical concepts that had no place in science. The question had to be resolved by seeking to understand the origins of life—this was the path followed by Oparin and Haldane. It was also to be resolved through the study of the simplest living organisms—which d'Hérelle and then the bacteriophage group attempted—which explains the importance research on viruses assumed in the 1930s and 1940s.[27]

Those involved in this work did not agree on what could be expected from it. For some, mainly chemists, loyal to a tradition which as we've seen went back at least to the nineteenth century, the question would be naturally resolved by a complete description of the components of life. For others, such as Delbrück and to some extent Schrödinger, it was a matter of revealing the physical principles and laws specific to the living world.

The Process of "Synthesis" in Science

The rise of the modern synthesis was not a redefinition of a field of knowledge because of new discoveries, but an ordering of a complex domain—that of the evolution of organisms—by assembling the knowledge and models from different disciplines. Such syntheses are not rare in science, but they are not very visible and have been studied less by historians than have, for example, scientific discoveries.

In an earlier chapter we encountered a transformation of biological knowledge that was also a synthesis—the emergence of cell theory in the middle of the nineteenth century. Cells had been known for more than two centuries, but cell theory put the cell and its nucleus at the center of the organization of the living world. Another synthesis, which we will describe in chapter 10, is the emergence at the beginning of the 1980s of the theory of oncogenes to explain cancer.

27. Podolsky, 1996.

Such syntheses are of interest because they reveal two aspects of scientific work that are often neglected in historical studies: the process of accumulation of knowledge, and the ongoing reorganization of that knowledge.

Contemporary Relevance

From Energy to Information

Since World War II, the idea of information has played a role equivalent to that of energy in the first half of the century. Information and its exchanges are everywhere, as much in the writing of molecular biologists as in that of ecologists and ethologists. And efforts to directly interpret the characteristics of living things and their evolution in informational terms are recurrent. The hope of reducing the complexity of phenomena to a single law, a single principle, is common in science and justified by the successes of the past. But there is always the risk, as seen in the description of energetics concepts above, of either being limited to vague analogies, or losing part of the complexity of the object under study during the process of reduction.

Current explanations through the existence of networks and the functional characteristics associated with them are also likely, if they aspire to supersede earlier explanations, to encounter the same difficulties.

From the Biosphere to Global Warming

As already mentioned, one of the interests of the history of science is in moderating the claims to novelty of some current concepts. A description of the biosphere and of the damage caused to the environment by human activity was provided in the 1930s. The existence of a new age, the Anthropocene, in which human activity has become the principal driver of global transformations, was asserted,[28] even if the term wasn't used. In contrast, what is very different from the current age is the reactions people had to these transformations. They appeared to be normal and unavoidable. They were seen as the continuation and amplification in the human species of tendencies that have been manifest throughout evolutionary history. The place of human beings at the head of evolution wasn't justified by religious considerations, but came naturally out of observations made. As Dobzhansky expressed particularly clearly, the duty of human beings was to embrace that position and, fortified with the knowledge they had gained, to take charge of that evolution: which meant replacing natural selection with a well-conceived eugenics policy, and not hesitating, when the state of knowledge allowed it, to biologically improve living species, including the human species.

The Responsibility of Biologists

We have referred several times to the interest shown by many biologists in eugenic ideas; the value of eugenic measures, as they might be applied, has been debated, but not the reasons for biologists' interest in eugenics policies.

Few biologists have been active participants in the establishment of eugenics policies,[29] but few have been opposed to them, either. It is not a matter of retrospectively judging the attitudes of the prewar years, but simply of attempting to understand what

28. Bonneuil and Fressoz, 2013.

29. Bashford and Levine, 2012.

prevented these biologists from seeing the dramatic consequences of the measures that were then implemented.

An initial, partial response is that there was overconfidence in the value of knowledge—or, more precisely, a complete unawareness of the extent of their ignorance. A second is that there was a shared objective that the acquired knowledge served. A justification of scientific work was all the more necessary when economic crises limited state resources, and when Marxism valued the work of scientists on the condition that it was oriented toward the good of society. A third element favorable to such blindness was the false assertion that all problems affecting society have a biological origin and should therefore be resolved with the use of biological knowledge. More specifically, a genetic problem, such as an accumulation of undesirable mutations, should be resolved through an intervention of a genetic nature—by controlling the transmission of allelic forms from generation to generation. Finally, eugenics was carried along by a blind faith in science and a quasi-messianic view that made scientists the new priests. The will to overcome religion's resistance to human intervention constituted a strong motivation.

Might analogous errors in judgement lead current work in directions that, in a few decades, will appear unethical? Aren't the proponents of gene therapy committing the same error eugenicists did in wanting to act directly on the genome to correct genetic defects, rather than seeking alternative solutions—for example, developing medications to counteract the metabolic disturbances caused by these genic alterations?

10

Twentieth–Twenty-First Centuries

AFTER THE SYNTHESES

The Facts

Progress in the life sciences has been so rapid over the past 50 years that it seems almost impossible to present a panoramic view of this recent history. The huge metabolic maps describing all the chemical reactions happening inside living organisms are testimony to the enormous progress that has been made since the deciphering of the first metabolic pathways and cycles.

After World War II, the sciences benefited from a great deal of state funding through both preexisting organizations and newly created ones, such as, for example, the National Institutes of Health and the National Science Foundation in the United States,[1] the Medical Research Council (for biology and medicine) in the United Kingdom, and the Centre national de la recherche scientifique in France. Such grants were boosted by

1. Appel, 2000.

the addition of private funding, such as that from the Rockefeller Foundation. Scientific research became a national priority to improve health, develop industry, and to confront the risks posed by the Cold War. Such funding continued after that war had ended.

The recent transformations of biology often rest on technological advances, such as the use of radioisotopes—a fallout from the American nuclear program—in biochemistry, medicine, and ecology,[2] or more recently the development of effective methods for sequencing DNA.

Other techniques, such as the inactivation of genes through homologous recombination (a technique known as "knockout"), or the use of RNA interference to selectively inhibit a gene's expression (through "knockdown"), have contributed significantly to the deciphering of molecular functions. We should also mention more "modest" techniques and tools that nonetheless played and still play a major role, both in basic research and in applied research. This is the case of nucleases capable of cutting DNA at precise positions. In the mid-1970s, the discovery and mastery of restriction enzymes played a major role in the rise of genetic engineering. More recently, the characterization of the CRISPR-Cas9 system, in which a nuclease (Cas9) is guided to its cutting site by an RNA, opened up new possibilities, in particular in gene therapy. This advance was the result of the encounter between the manufacture of nucleases to cut at precise locations in the genome (ongoing for close to 30 years), and the unexpected discovery of a bacterial defense mechanism against viruses (CRISPR), which uses the nuclease Cas9. It paved the way to the editing of the genome—that is, to

2. Creager, 2013.

targeted modification aiming in particular to eliminate errors resulting from mutation. Would the new system prove to be reliable and effective enough to allow a rapid expansion in gene therapy, whose role today, despite much effort, remains limited? Will it be possible to go beyond it, and to reliably transmit the modifications thus introduced into the genome to future generations?

To simplify things, we will group the developments of the past few decades along five axes, some of which are obvious—such as the neurosciences and the sequencing of genomes—and others less so, though equally important, such as the progress in structural biology with which we will open this chapter.

The Rise of Structural Biology

At the beginning of the 1960s, when molecular biology was becoming a dominant discipline, the structures of only two proteins, myoglobin and hemoglobin, had been described with any precision. The structures of enzymes were described at the end of the 1960s and, from that, mechanisms explaining how those enzymes catalyzed reactions were proposed.

The elucidation of structures has continued to accelerate up to the present. This acceleration is the result of the convergence of multiple advances: an equally exponential increase in the processing power of computers; an increase in the energy of X-ray beams used in diffraction experiments (with synchrotron radiation produced by particle accelerators); the preparation of pure proteins, potentially modified or marked, through genetic engineering; and the automation of all the stages of the process, from choices in the conditions of crystallization to the representation of structures on the computer screen.

At the beginning of the 1980s, the determination of structures led to the general adoption of a new way of representing proteins, the so-called "ribbon" diagram, highlighting the secondary structural elements within the protein structure. The development of this representation was the work of a single person, Jane Richardson (1941–).[3] Trained in philosophy, she was convinced of the importance of representations in science, and she devoted more than two years of her scientific career to this project. This ribbon diagram proved to be particularly useful for the comparison and classification of protein structures. It also proved to be well adapted for the assimilation of proteins into nanomachines, because it made it easy to see the relative movements of the different parts of proteins during their functioning. Despite the importance of this representation, Jane Richardson's name and her contribution are little known—yet another example of an accomplishment essential to the development of the sciences that escapes the usual channels of recognition.

The structures of increasingly large protein assemblages—viruses, ribosomes—have been elucidated using a combination of X-ray diffraction and very high-resolution electron microscopy. The structures of a class of proteins that had hitherto remained resistant to crystallization, membrane proteins, were gradually revealed from the mid-1980s. Membrane proteins (and their assemblages) play essential roles in cells: they capture light energy in photosynthetic organisms; produce cell energy in the form of ATP; are receptors for hormones, growth factors and neurotransmitters; and act as channels enabling the selective passage of ions—a particularly important function in nerve cells.[4]

3. Morange, 2011a.

4. Trumpler, 1997.

As soon as the first protein structures were described, their presumed limits were pointed out. As they had been fixed by the process of crystallization, critics said, they were unlike the dynamic structures that proteins had in solution, and only the latter would able to account for protein function. Other techniques, such as molecular modeling and nuclear magnetic resonance, would need to be developed, allowing access to molecular dynamics.

Between 1960 and 2000, such criticism had little influence. The development of the new techniques was difficult, whereas the fixed structure revealed through X-ray diffraction, on the contrary, proved eminently suitable for explaining the functioning of proteins. From the early 2000s, the landscape was rapidly transformed. New techniques finally enabled a description of the extraordinary fluidity of protein structures, inaccessible until then. The allosteric model, developed in 1965 by Monod, Jeffries Wyman (1901–1995), and Jean-Pierre Changeux (1936–), by which proteins exist as an equilibrium of multiple conformations,[5] until then reserved for a few regulatory proteins, was broadened to include all protein structures (as well as nucleic acids). This new dynamic view was consistent with the recent observation that many proteins are only partially folded. It also introduced the possibility of finding new therapeutic agents that would not impact the active sites of proteins, but would alter the balance between different protein states. It offered possible scenarios for the acquisition of functions by the first proteins formed at the beginning of life, as well as for their later evolution. From the mid-1990s, the study of proteins and of individual protein complexes through physical techniques fit perfectly with this new dynamic description of macromolecules.

5. Monod et al., 1965.

This dynamic approach was not limited to isolated proteins. Other technological developments—such as the production of fluorescent hybrid proteins (obtained through coupling with GFP, a fluorescent protein extracted from jellyfish)—enabled the movements of proteins inside living cells to be tracked. This revealed the importance of stochastic phenomena at the molecular and cellular level. At this scale, and considering the small number of macromolecules involved, all molecular processes—such as the binding of proteins to DNA—are stochastic. The level of expression of a gene represents only a mean value, as its expression fluctuates both in time and from one cell to another. The role of this stochasticity, this "molecular noise" as specialists call it, is an open question. Is it a source of variability in living things? Or is it the cause of disturbances against which organisms have learned to fight?

Advances in the knowledge of structures and, more recently, in the dynamics of molecular events have led to a profound reconception of cellular and physiological phenomena. The traditional description of the different phases of cell division, mitosis and meiosis, developed at the end of the nineteenth century, has been replaced by a precise description of the proteins that participate in it, validated by observations of living cells. The signaling pathways that enable hormones and growth factors to affect target cells were deciphered at the beginning of the 1980s. At the same time, the many proteins that are involved in the control of gene expression were described, as were their complex interactions. Since the late 1990s, a great many small RNAs—such as microRNAs—have been discovered; they are also involved in the control of gene expression. The movement of proteins, via a complex network of vesicles, from their place of synthesis to the structure (organelle, membrane) where they carry out their functions, has been gradually deciphered over the last three decades. The molecular steps of processes as complex as cellular respiration and photosynthesis have been elucidated.[6]

This progress in the precise structural description of molecular components did not come with a loss of the power of cells. The reductionist tendencies of the 1960s quickly gave way in the 1970s to a renewed interest in the structures and functions of the cell. In 1972, Seymour Singer (1924–2017) and Garth Nicolson (1943–) proposed a new model for the structure of cell membranes. Until then, very little was known about the cell membrane. In the second half of the nineteenth century, many biologists had even denied its existence.[7] In the early 1960s, Peter Mitchell (1920–1992) proposed a mechanism to explain the production of energy through redox reactions occurring inside mitochondria, attributing a major role to the internal membrane of the mitochondrion. This mechanism was finally accepted at the end of the 1970s. A similar model enabled an explanation of photosynthesis, the capturing of light energy by chloroplasts, organelles present in plant cells. The reactions that use the energy of light to fix carbon had been deciphered in the late 1940s by Melvin Calvin (1911–1997).

The concept of stem cells emerged in the late 1950s and early 1960s from work on the renewal of blood cells, at the same time as the characterization of ovarian and testicular tumors in mice. The therapeutic potential suggested by the production of human embryonic stem cells in 1998 explains the enormous amount of research that is devoted to them today.

Two examples illustrate this transformation of knowledge: that of a discipline, immunology,[8] and that of a disease, cancer.

6. Govindjee et al., 2005.

7. Lombard, 2014.

8. Nagy, 2014.

Progress in the biochemical characterization of antibodies, parallel with the development of molecular biology, showed that the great diversity of antibodies cannot be attributed to a different folding of identical molecules, as Pauling had hypothesized in the 1940s. The diversity is of genetic origin; however, the number of genes required to engender such diversity did not seem compatible with the known size of genomes. This paradox was solved at the end of the 1950s with the model of clonal selection proposed by Niels Jerne (1911–1994) and Frank Macfarlane Burnet (1899–1985). Cells manufacturing antibodies of different specificities exist before invasion by any external agent. In the late 1970s, Susumu Tonegawa (1939–), using the new tools of genetic engineering, showed that genes encoding these antibodies resulted from a very complex process involving a recombination of genetic fragments during the formation of the B lymphocytes that manufacture antibodies, a process that leads to the extraordinary diversity of the cells' products.

Before the details of these mechanisms were described, in 1975 Georges Köhler (1946–1995) and César Milstein (1927–2002) had shown that it was possible to obtain cell lines that produced unlimited quantities of an antibody specific for a well-defined structure, through the fusion of an antibody-producing cell and a cancer cell. These monoclonal antibodies proved to be an extraordinary tool for basic research and an effective therapeutic agent in the fight against cancer. The production of monoclonal antibodies was the only important outcome of the work on cell hybridization carried out from the 1960s, work which many scientists had hoped would allow genetic research to move from the level of the organism to that of cells.

In the early 1960s, the Australian immunologist Jacques Miller (1931–) demonstrated the role of the thymus in the development of the immune response. A second type of lymphocyte—the T lymphocyte, or T cell—was described; T cells control the activity of B cells. In the thymus, T lymphocytes that recognize components of the self are sorted and eliminated. T cells also act directly on their targets—"foreign" cells such as, for example, the cells of a transplant, or cells infected by a virus—to destroy them. Genes coding for receptors present on the surface of T cells that recognize the targeted cells are also the result of a process of genetic fragment recombination, which is the origin of their diversity.

Since this time, during which humoral immunity and cellular immunity were put back on equal footing, immunologists have devoted a great deal of work to describing the complex interactions between B lymphocytes and T lymphocytes, but also the many other cell types participating in the immune response (macrophages, Langerhans cells, regulatory T lymphocytes, NK [natural killer] cells) and the signals that these cells exchange. The idea of an immune network proposed by Niels Jerne in 1974 remains valid, but not in the precise form that he gave it.

However, it should be recognized that although this knowledge enabled progress in organ transplantation—thanks, in particular, to the discovery of molecules inhibiting the immune response, such as cyclosporin, used in the late 1970s—the immune response as a whole, and especially its dysfunctions and failures, is still poorly understood. The origin of autoimmune diseases remains a mystery. The cause of the current increase in allergies remains unknown. And the difficulties encountered in developing an effective vaccine against HIV, the causative agent of AIDS, show the limits of our knowledge today. These difficulties can be contrasted with the rapid development, thanks to what we have learned about the structure of the HIV virus, of effective medications to treat AIDS, in particular protease inhibitors aimed at one of its enzymes, which have completely transformed the prognosis of the disease.

This transformation in conception and, to a lesser extent, practice, is also evident in the case of cancer. Since the nineteenth century and the work of Virchow and his students, the search for an explanation for cancer has always remained closely associated with advances in biological knowledge. In the 1970s, various different explanations for cancer coexisted, none being obviously superior to the others. The theory that cancer is the result of a "chromosomal aberration," suggested by Theodor Boveri at the beginning of the twentieth century, has just been directly confirmed for certain forms of leukemia. Another theory was that cancer is the result of "somatic" mutations occurring throughout life, but an obstacle to this theory was the existence of apparently nonmutagenic carcinogenic agents. In 1973, that obstacle was overcome by Bruce Ames (1928–), who demonstrated that certain carcinogenic substances become mutagenic following their metabolic transformation within an organism. Since cancer cells often have characteristics that resemble those of embryonic cells, many biologists thought that cancer was the result of an alteration of the mechanisms regulating gene expression in differentiated cells. Cancer could also be the result of a viral infection.

Two distinct lines of research in the early 1980s led to the development of a new explanatory model for cancer.[9] The characterization of oncogenic viruses, so called because they are capable of transforming cells and forming of tumors, by Harold Varmus (1939–) and Michael Bishop (1936–) and their teams, revealed that the viral genes responsible were cellular genes that had been captured and modified by the viruses. At

9. Morange, 1993.

the same time, Robert Weinberg (1942–), using the same experimental approach Avery had used 40 years earlier, showed that normal cells could be transformed into cancer cells through the addition of DNA fragments from cancer cells. The genes carried by those fragments were shown to be identical to those captured by oncogenic viruses. Their sequencing revealed that they coded for cell receptors, and more generally for components of signaling pathways involved in the control of cell division. The mutations leading to cancer were ones that activated these genes. In 1986, Robert Weinberg and his collaborators isolated the first gene believed to be a tumor suppressor—whose inactivation, in contrast, led to the formation of a tumor—the gene responsible for retinoblastoma, a cancer of the eye. Oncogenes and tumor suppressor genes work together in the formation of cancers.

The development of this new view of cancer was made possible through the use of the tools of genetic engineering. The new view reconciled earlier explanations: cancer can be the result of spontaneous mutation, the action of a virus, or chromosomal rearrangements—leading to the activation or deactivation of genes—and corresponded to a cellular dysregulation. The modification of signaling pathways is at the heart of cancerous transformation, and a description of these pathways has been accomplished alongside a description of their alteration in cancer.

The search for oncogenes and tumor suppressor genes was one of the reasons for undertaking the complete sequencing of the human genome, a project launched in 1986. Knowledge of the structures of proteins encoded by these oncogenes has paved the way for the manufacture of "magic bullets" (to use Ehrlich's expression), chemical molecules or monoclonal antibodies capable of specifically inhibiting the proteins' activity.

Forty years have gone by. Such therapeutic agents have been developed and their effectiveness demonstrated; and yet the mortality rate associated with cancer has continued to rise. Changes to the environment, an increase in life expectancy, and the successes in the fight against other afflictions, such as cardiovascular disease, in part explain that apparent failure. But the new model and its applications have also had to confront their own difficulties. The new therapeutic agents have proved to be effective, but only on certain forms of cancer, and phenomena of resistance have appeared. The number of oncogenes and tumor suppressor genes has continued to grow, as has the number of mutations necessary for the formation of a cancer cell (from six to eight, depending on the model). Not only are different genes mutated in different types of cancer, but each tumor is a unique case. A tumor is itself heterogeneous, and the cells forming it do not carry the same mutations; they compete during the development of the tumor in what resembles an evolutionary process. This has led to the notion that a precise molecular knowledge of every tumor (through sequencing) is necessary in order to choose the best treatment to fight it. Each cancer patient would then receive a personalized treatment.

The difficulties encountered in the development of new anticancer medicines are part of a much larger phenomenon: a reduction in the number of new active pharmacological agents placed on the market. This problem can be seen particularly well in the case of antibiotics. It is explained in part by the organization of the pharmaceutical industry, as well as by an increase in the regulations accompanying the release of a new drug onto the market. But there are still fundamental unanswered questions about the very foundations of the project Ehrlich undertook more than a century ago. The strategy of the rational "design" of new medications and that of systematic "blind screening" in chemical libraries of molecules have clashed, without one approach winning over the other. More seriously, it seems as though most of the active principles have already been described, as though the space for active molecules is now a closed, rather than an open, space.

The Encounter between Molecular Biology and the Modern Synthesis

Evolutionary theory has been considerably enriched since the rise of population genetics and the modern synthesis.[10] Many models have been constructed to explain evolutionary facts whose existence often seemed counterintuitive. A good example of this is the so-called "Red Queen" model proposed by the American evolutionist Leigh Van Valen (1935–2010). The name was borrowed from a passage in Lewis Carroll's book *Through the Looking-Glass*. During her race with the Red Queen, Alice is surprised when the landscape doesn't change, and the Queen responds that they are running precisely in order to stay in the same place. The model applies well to organisms that are in close interaction, such as a virus and its host. The equilibrium maintained between the two is the result of an arms race, each evolving rapidly to maintain the balance of terror. And this model is directly applicable in the domain of epidemiology.

A question that has been of great interest to evolutionary scientists is that of the level where selection occurs. The attempt by Vero Copner Wynne-Edwards (1906–1997) at the beginning of the 1950s to demonstrate the existence of supraindividual selection on the basis of group behavior (involving reproduction, for example) came up against the contemporaneous, but

10. Heams et al., 2009.

opposite, movement asserting a selection descended to the level of the gene.[11] This issue was the focus of the 1966 book *Adaptation and Natural Selection* by George C. Williams (1926–2010),[12] and then of the work of William Hamilton (1936–2000) on kinship selection, continuing that of Haldane and Fisher. The aim was to explain the existence of altruism in nature, as proposed in 1902 by the Russian anarchist Petr Kropotkin (1842–1921), despite such behavior being a priori incompatible with the theory of natural selection.[13] If the only thing that matters is "fitness"—that is, the number of descendants that an individual produces—how can it be explained that an organism might sacrifice itself for its fellow creatures? Hamilton was the first to propose a solution: an organism would sacrifice itself if, even without leaving descendants, its behavior would contribute to transmitting its genetic forms to future generations, by promoting the reproduction of those—its relatives—that carry the same gene forms as its own. Robert Trivers (1943–) demonstrated that reciprocal altruism—for example, birds taking turns performing as lookout—could benefit the reproduction of all, and John Maynard Smith (1900–2004) removed the obstacle represented by the actions of "cheaters." Richard Dawkins (1941–), using many nicely chosen metaphors, shed a great deal of light on this in *The Selfish Gene*, published in 1976.[14]

However, the debate had not been settled. Relaunched by Gould, it gradually evolved into a consensus that selection operates on many levels, but the gene and the organism are its principal levels of action.

11. For the former, see Wynne-Edwards (1962) and Borrello (2008).

12. Williams, 1966.

13. Dugatkin, 2011.

14. Dawkins, 1976.

This promotion of the gene as the level of selection was an indirect consequence of the growing importance of molecular biology. In the 1960s, however, this came up against models of the modern synthesis.

Richard C. Lewontin (1929–) and John Lee "Jack" Hubby (1932–1996) launched the attack by demonstrating that level of allelic diversity as estimated by techniques of immunology and electrophoresis was far above that commonly accepted by geneticists. At the same time, the comparison of protein sequences from different species, initiated by Émile Zuckerkandl (1922–2013) and Pauling,[15] revealed a constant rate of mutation, converting the measurement of the divergence of sequences into a molecular clock, a result seemingly incompatible with the irregular sifting of those mutations through natural selection.[16] The Japanese geneticist Motoo Kimura (1924–1994) would expand these observations by inferring that most variations are neutral at the molecular level, without phenotypic effects likely to be acted upon by natural selection.

The neutral theory of molecular evolution today occupies an important place in the models of evolutionary scientists. The geneticist Michael Lynch (1951–) proposed that the increase in genome size observed in multicellular organisms is the result of neutral mutations not sifted out by natural selection.[17]

In the 1970s, Gould also challenged, in different ways, the major role of natural selection.[18] In 1979, in an article on the spandrels of San Marco cowritten with Lewontin, Gould suggested that many of the traits of organisms are explained not by

15. Zuckerkandl and Pauling, 1965.
16. Dietrich, 1994; Suarez and Barahona, 1996.
17. Lynch, 2007.
18. Sepkoski, 2008.

the optimizing action—"Panglossian," as he put it—of natural selection, but by constraints that exist during development; just as the spandrels of the Saint Mark's Basilica in Venice were not created to house the images of the four evangelists, but were necessary to raise the cupola.[19] The fact that a structure present in an organism might be subject to the action of natural selection didn't mean that its existence was the fruit of that action. He introduced the term "exaptation" for these later adaptations of preexisting structures.[20] The existence of an irregular evolution of organisms—with long periods of stasis interspersed with rapid variation, a concept which in 1972, along with Niles Eldredge (1943–), Gould named "punctuated equilibrium"—he believed to reveal those constraints restricting the action of natural selection.[21] Catastrophic events, such as the fall of a meteorite, could also profoundly alter the evolution of the living world, independently of the adaptive capabilities of the organisms that survived, or those that disappeared. Gould's work was also an occasion to give paleontology, which had become paleobiology, a second wind, through the application of statistical methods begun by George Simpson, and thanks to the introduction of modeling by ecologists.[22]

Following the work of Jacob and Monod, the distinction made by molecular biologists between structural and regulatory genes led many biologists to see variations in the latter as the motor of evolution. In 1977, in his work *Ontogeny and Phylogeny*, Gould critiqued Haeckel's theory of recapitulation, but suggested that some of the observations of Haeckel and his disciples

19. Gould and Lewontin, 1979.
20. Gould and Vrba, 1982.
21. Gould and Eldredge, 1977.
22. Sepkoski, 2012.

could be explained by the existence of mutations affecting the rhythm of embryonic development, or heterochronic mutations.[23] In 1975, Mary-Claire King (1946–) and Allan Wilson (1934–1991) demonstrated the short genetic distance separating the human being from the chimpanzee, and from that deduced that the genetic mutations responsible for the greatest differences between these two species involved a small number of regulatory genes that had not yet been characterized.[24]

In the early 1960s, the geneticist Antonio García-Bellido (1936–) characterized the role of selector genes, which he identified as regulatory genes, in the morphogenesis of fruit flies. During the 1970s, a genetic framework for the embryonic development of the fruit fly was built and widely disseminated by Crick and Peter Lawrence (1941–).[25]

This new view of fruit fly development was supported by observations that geneticists had been making since the 1910s, with the characterization of many mutations affecting embryonic development, and in particular the work by Edward B. Lewis (1918–2004) on the gene complexes *bithorax* and *Antennapedia* and the homeotic transformations that their mutations induced—a transformation of one part of the organism into another. Experiments carried out in the late 1970s and the early 1980s strengthened the new developmental model. In 1978, Christiane Nüsslein-Volhard (1942–) and Eric Wieschaus (1947–) described all the genes involved in the early embryonic development of the fruit fly, which they grouped according to their order of intervention and their effects in embryogenesis. Beginning in 1984, the mutations of genes responsible for

23. Gould, 1977.

24. King and Wilson, 1975.

25. Crick and Lawrence, 1975.

homeotic variations were characterized. A strong evolutionary conservation of the function of these developmental genes was demonstrated. In 1995, Walter Gehring (1939–2014) showed that the same gene, *Pax6*, controlled the formation of the eye of a mammal and of a fruit fly.[26]

These unexpected results led to the rapid emergence of a new discipline, evolutionary development, or evo-devo, at the interface between evolutionary biology and developmental biology.[27] For some converts to the new discipline, evolutionary transformations were seen as the consequence of structural modifications or changes in the regulation of developmental genes: describing these modifications was considered to be the primary objective of evolutionary science.

The development of evo-devo constituted a double break with earlier practices and models. The first concerned the rapid diversification of model systems. While molecular biologists have often been content with a single model—bacteria—to research the fundamental principles of life, evo-devo specialists are interested in the most diverse and often the most exotic organisms, and encourage the sequencing of their genomes.

The second break was with the dogmas of the modern synthesis: all genes are not equal, and variation in developmental genes plays a major role in evolutionary transformations. Nor are all mutations of the same nature: for Sean B. Carroll (1960–), one of the leaders of this new domain, evolutionary variations are in most cases modifications of sequences upstream of genes that control their expression.[28] These are the only kinds of mutation that enable the transformation of organisms—for ex-

26. Gehring and Ikeo, 1999.

27. Laubichler and Maienschein, 2007.

28. Carroll, 2008.

ample, causing the expression of a gene in a tissue where it hadn't previously been expressed, without altering its overall function: in a way, the kind of transformation that Cuvier would have accepted!

Some evo-devo specialists, such as Eric H. Davidson (1937–2015), have gone even further in challenging the modern synthesis.[29] After having organized the genes participating in the development of the sea urchin embryo into a complex network, Davidson divided them into central and peripheral groups. The mutation of genes in the former group leads to extremely significant variations, at the origin of new phyla. In contrast, variation of genes in the peripheral group leads to the formation of new species or new genera. Thus, it is the nature of mutations, and not evolutionary history, that determines the shape of the evolutionary tree. Evolutionary history is discontinuous, with the sudden emergence of organisms with new characteristics, the equivalent of Goldschmidt's "hopeful monsters."

However, it isn't simply a matter of returning to concepts that predate the modern synthesis. The new hypotheses are experimentally testable using "synthetic experimental evolution," which involves introducing into a model organism a genetic variation that is thought to play a major role in evolution, and observing the effects of its addition on embryonic development.[30]

The meeting of molecular biology and evolutionary theory cannot be reduced to a description of research in the realm of evo-devo. Symbiosis, the close interaction of two organisms, had been demonstrated in the plant world in 1866 by the Swiss biologist Simon Schwendener working with lichens—a symbiosis between a fungus and an alga. The role of symbiotic microorganisms in

29. Davidson, 2006.

30. Erwin and Davidson, 2009.

nitrogen fixation by plants was also already known. The hypothesis, advanced at the end of the nineteenth century but forgotten and proposed anew in 1967 by Lynn Margulis (1938–2011), that eukaryotic cells are the result of symbiotic events—mitochondria being the remnants of ancient bacteria "swallowed" by other cells—has been amply confirmed since then, thanks to molecular data.[31] Similarly, molecular tools have enabled the demonstration that genes can be transferred horizontally, from organism to organism, and not just vertically through heredity. These results provide two new mechanisms of genetic variation.

In a less visible but more important way, the encounter between functional biology and evolutionary biology has involved many fields of research.[32] For example, ethologists, geneticists, and neurobiologists interact ever more closely to explain the behavior of animals and their evolution. The technology of protein engineering has also made great advances. When, at the beginning of the 1980s, the development of techniques for directed mutagenesis paved the way for the modification of proteins, the first attempts aimed to manufacture enzymes with modified activity or specificity by using a "rational" approach: changing a few well-chosen amino acids, generally located in the active site of the enzyme.

The first attempts were positive, but their success was nonetheless limited. The enzymes thus modified proved to be either unstable or of low activity. Targeted modification has been supplemented with (or sometimes even substituted by) a step of "directed" evolution, by random mutagenesis and selection of the sought-after properties. The same approach has been

31. Sapp, 1994.
32. Morange, 2005.

adopted for the manufacture of new genetic circuits in synthetic biology. These new procedures are a recognition of both the difficulty of accounting for complex parameters, such as the stability of proteins and their ability to change conformation during the act of catalysis, and the creative power of natural selection.

Little by little, the evolutionary history of proteins is unfolding. The comparison of the sequences and structures of homologous proteins from different species has enabled the reconstruction of the ancestral form of those proteins, and the steps that have led to the current forms. Combined with in vitro evolutionary experiments, such as those begun more than 30 years ago by Richard Lenski (1956–) on bacteria, these experiments have contributed to making these nanomolecular machines of life not only superb objects of study for physicists, but also the products of an evolutionary history gradually unveiled. This is the best answer to the proponents of "intelligent design," to whom the complexity of proteins and enzymes appears the perfect example of something the current theory of evolution cannot explain. Evolutionary biology and functional biology have come together to form a new "functional synthesis."

Epigenetics is a strange thing. Since the early 2000s, it has been very prominent in research, but also in the media.[33] Epigenetic modification—DNA methylation and the many modifications of the histones that surround DNA—plays an important role in the control of gene expression during cellular differentiation. It is also implicated in certain pathologies—cancer, psychiatric illnesses—and, more generally, in changes that occur during the aging process. It can occur in response to signals from the environment and it might, in some cases, be

33. Jablonka and Lamb, 2005.

transmitted to descendants, which would open the door, as we have seen, to a return to the inheritance of acquired traits.

There is a certain vagueness that surrounds the term "epigenetics," which can refer to the precise mechanisms of regulation described above but also any biological phenomenon that can be modified by the environment and which, for that reason, escapes a strict genetic determinism.

The complex history of the term "epigenetics" explains this polysemy. It has been introduced at least five times to designate five different types of phenomenon. At the beginning of the twentieth century, embryologists called embryonic processes that seemed independent of genes and chromosomes—whose existence was considered a type of preformationism—epigenetic (the adjective corresponding to epigenesis). This use would persist for the entire century; for example, the establishment of synapses between neurons was considered an epigenetic phenomenon because it depended on the early functioning of neurons and not on the instructions contained in the genome. In 1942, Waddington, often considered the father of epigenetics, introduced the term "epigenetic"—a combination of the words "epigenesis" and "genetics"—to designate the science, which didn't yet exist at that time, that would describe the role of genes in embryonic development and cellular differentiation.[34] In the 1950s, many laboratories studied phenomena of epigenetic heredity—that is, phenomena of heredity not carried by genes. In 1958, David L. Nanney (1925–2016) distinguished two genic functions: first, to determine the precise structure of a cell component, and second—the function he called epigenetic—to precisely regulate the production of that component in the time and space of development. Epigenetic

34. Waddington, 1942.

marks were gradually described in the 1960s and 1970s, and their roles gradually elucidated.

These various meanings have often been combined; however, the fact that a gene is regulated by epigenetic modifications does not mean that the modifications thus characterized are transmitted to descendants, nor that they are independent of genes.

One must contrast the polysemy of the term "epigenetic" and the fragmentary nature of the experimental data concerning hereditary transmission of these marks—at least in the animal world—with the effects of the popularity of epigenetics, and the sometimes overblown hopes placed in its development. It has led to aspirations of escaping genetic determinism, to be able, through our behavior, to "master" our genes; also the hope of explaining the genesis of certain illnesses (and fighting against them)—such as autism or schizophrenia—for which genetic explanations are diluted in the vast number of genes involved. Epigenetics is a sort of post-genetics, a reflection of the difficulties encountered in genetics. It is the heir of the vast amount of work carried out during the twentieth century to counter the dominant role of the nucleus and of chromosomes in hereditary phenomena: from the description of cytoplasmic heredity by Carl Correns, and the work of Boris Ephrussi and Piotr Słonimski (1922–2009) on yeast, which would lead to the discovery of mitochondrial DNA, to the demonstration of the transmission of the shape of the cortex in paramecia during cell division by Janine Beisson (1931–2020) and Tracy Sonneborn (1905–1981).[35] It is also proof that the general public, but also many biologists, have never fully accepted the place granted to genes, or the inexistence of an inheritance of acquired traits.

35. Sapp, 1987.

Genome Sequencing

After the development of DNA sequencing in the mid-1970s, the first targets were viruses.[36] The decision to sequence the entire human genome was made in 1986. There were multiple reasons for this decision, ranging from the need to use up left-over research funds to plans to characterize all genes (onco-genes) involved in the formation of cancers. And it had become obvious, since the structure of DNA had been determined and the genetic code deciphered, that one day it would be essential to sequence the human genome.

There were many debates on the need (or not) to sequence noncoding DNA, and on the interest of simultaneously sequencing the genomes of other organisms to compare them with the human genome. The first genome sequenced was that of a bacterium (1995), followed shortly by that of a yeast (1996). The sequencing of the human genome was declared complete in 2001, earlier than had been predicted. The sequencing was carried out by using a technique developed by Frederick Sanger in 1977, without the technological revolution that some had considered necessary to undertake the project. Computers were increasingly used to assemble the sequences, to the point that the "shotgun" technique—random fragmentation of the genome, sequencing the fragments, then assembling them—advocated by Craig Venter (1946–), widely replaced the mapping approach developed by the International Human Genome Sequencing Consortium.

The results were sometimes considered disappointing, out of step with the hopes that many had held out. Sequencing was more a tool for future work than a result in itself. However, it revealed a

36. Garcia-Sancho, 2012; Stevens, 2013.

certain number of facts that dramatically changed the way biologists viewed their work. The human genome contains a small number of genes, comparable to the number present in the fruit fly, and fewer than the fairly simple plant *Arabidopsis thaliana*. The genome also proved to be very rich in parasitic sequences that were repeated in multiple copies. Most of these sequences are, or more often used to be, transposons—mobile genetic elements capable of moving within the genome, first described by the American geneticist Barbara McClintock in the 1950s, and whose characteristics were specified in the 1970s, in particular in bacteria.[37] The recent accumulation of genomic sequences has provided an evolutionary history of genomes—with partial or entire duplications, rearrangements, and so on—alongside an evolutionary history of genes. It enriches the study of genes and their mutations without making that study obsolete.

The New Frontier: The Neurosciences

The molecular approach that was established in the 1960s had an immediate, though not necessarily fortunate, impact on neurobiology: an explanation of neural phenomena was to be sought directly at the molecular level. Many neurobiologists attempted to show that behaviors were stored in the form of molecular memory.[38] Their results disappeared from the scientific literature in the mid-1970s to make way for a less clearly reductionist approach to neural phenomena.

The term "neurosciences" began to be widely used in the 1960s. The experiments on cats conducted at Harvard beginning in

37. For McClintock, see Comfort, 2001.

38. Irwin, 2007.

1958 by David H. Hubel (1926–2013) and Torsten N. Wiesel (1924–) played an important role in the promotion of research on the brain. Through electrophysiological recordings of neurons in the visual cortex, they showed that information was "preprocessed" there: some neurons responded only to certain geometric characteristics of detected objects (for example, different orientations) or to movements. They also showed that vision is established during a critical phase of brain development. These observations are at the origin of the ever-growing interest in neural plasticity.

In 1981, the Nobel committee awarded its prize to Hubel and Wiesel, but also to Roger Sperry (1913–1994), whose experiments demonstrated that the nervous system is also wired in a rigid and irreversible way.[39] By studying patients whose corpus callosum—the fibers linking the two cerebral hemispheres—had been severed to prevent an increase in epileptic seizures, he showed that the two hemispheres deal with information in parallel and independently. To explain the formation of a nervous system wired with such precision, he hypothesized that the cells and the fibers of the brain carry unique identification markers that enable the correct addressing of axons during embryonic development, a hypothesis that was later confirmed molecularly.

This return, in a more sophisticated form, of cerebral localization was not limited to the neocortex. The centers that control the various phases of sleep or engender a feeling of pleasure were also localized between 1960 and 1970, also using a combination of recording and electrical stimulation, and targeted destruction in animal models. Sometimes, the neurons involved use a specific neurotransmitter, which has led to the neurotransmitter being too narrowly associated with a particular behavior.

39. Horder, 2008.

At the beginning of the 1980s, the rise of the neurosciences benefited greatly from techniques developed by molecular biologists. The structures of neurotransmitter receptors and of ion channels,[40] and the molecular mechanisms that control the release of neurotransmitters, were described in rapid succession. The neurosciences also benefited from progress in electrophysiology, and in particular from the development of the "patch clamp" technique, which, with the use of a microelectrode, enabled the characterization of the electrical properties of isolated ion channels. With the techniques of immunocytochemistry and of in situ molecular hybridization, the different types of neurotransmitters and receptors were localized in the nervous system. The classical anatomy of the brain was gradually superseded by a molecular anatomic model.

As early as the mid-1950s, anatomopathological observations had revealed the importance of a specific cerebral structure, the hippocampus, in processes of forming memories (the structure that Owen erroneously claimed was possessed only by humans). Soon after, electrophysiological recordings showed the existence of a slow neuronal response in this structure—the LTP, or long-term potentiation—which is associated with processes of learning and memory.[41] Using a simple biological model, *Aplysia*, the "sea slug," Eric Kandel (1929–) combined electrophysiological observations and an analysis of biochemical modifications that occurred simultaneously inside neurons to propose a molecular model of memory storage, breaking with 1960s model of molecular memory. The observed biochemical modifications, and in particular the activation of protein kinases, lead to a modification of synaptic contacts between neurons. In the 1990s, using trans-

40. Trumpler, 1997.
41. Craver, 2003; Craver and Darden, 2013.

genic mice, Kandel showed that similar biochemical phenomena occur in the neurons of the hippocampus during learning.

The neurosciences also benefited from the participation of many molecular biologists—Delbrück, Crick, Changeux, Gerald M. Edelman (1929–2014), among others—who saw a new frontier of biological knowledge in understanding how the nervous system functioned, one that would take the place of the last frontier, the description of genes, their role in hereditary phenomena, and the mechanism of their activity.

The development of the neurosciences was situated within a broader framework, that of the "cognitive sciences," a term introduced in 1973. The cognitive sciences were the creation of specialists in artificial intelligence, the heirs of the founders of cybernetics (Norbert Wiener [1894–1964], Warren McCulloch [1898–1969], and Walter Pitts [1923–1969]) and of computer science (Alan Turing [1912–1954] and John von Neumann [1903–1957]); they were joined by linguists such as Noam Chomsky (1928–) and philosophers (John R. Searle [1932–], Jerry A. Fodor [1935–2017], and Patricia S. [1943–] and Paul M. Churchland [1942–]).

The reductionist approach of many neurobiologists didn't work easily within this multidisciplinary group. The development of neuroimaging narrowed the gap between neurophysiological studies and psychological and behavioral approaches. The first method developed, positron emission tomography (PET), used radioactive metabolite analogues to localize active zones of the brain. More recently, functional magnetic resonance imaging (fMRI) has become the method of choice. Less invasive, it enables an estimation of the amount of oxygen carried by the blood to the different areas of the brain. And many other techniques are being developed.

To observe the brain while it is functioning, Broca's dream, finally seems to have become a reality. Beyond the confirmation (or the invalidation) of earlier cerebral localizations, the new techniques make it possible to search for a functional localization of a wider and more diverse range of tasks, and can prove functional associations between different areas of the brain—a means, perhaps, of gaining access to the ever-enigmatic "consciousness."

Optogenetics is the most developed form of the convergence between molecular biology, electrophysiology, and behavioral biology. By genetic engineering of neurons to make them express light-sensitive bacterial proteins, it became possible in the early 2000s for researchers to activate or inhibit neurons at will. By controlling the illumination, they are able to observe the effects of the change of protein activity in neurons on the behavior of model organisms: nematode, fruit fly, or mouse. Beyond the spectacular aspect of controlling the behavior of an organism from a distance, this new tool is above all a means to confirm, and potentially to correct, our knowledge of the role of the various neural circuits.

From the 1960s, a new medical definition of death was gradually introduced into the laws of many countries. For human beings, death became a cerebral death, confirmed by the technology neurobiologists used to study brain activity. If the primary motivation behind this legal reform was to enable organ transplants, the collateral, but no less significant, result was the recognition of the true value of the knowledge acquired by neurobiologists. Will this knowledge help overcome what even today seem to be insurmountable obstacles: the integration of observations made at various levels in the organization of the nervous system, and a scientific explanation of consciousness? Many biologists had hoped these obstacles would be overcome in the 1970s, but their hopes were dashed. Do today's hopes rest on more solid foundations? Only the future will tell.

A New View of the Living World

For its founders, the objective of molecular biology was to describe the fundamental mechanisms of life. In the mid-1970s, the development of genetic engineering confirmed that the level of knowledge attained was sufficient to be able to modify living organisms at will. Synthetic biology, which appeared at the beginning of the 2000s, was a continuation of this; it differs from genetic engineering in its systematic approach (in the engineering spirit) and loftier ambitions—going so far as to attempt to create completely artificial organisms. The complete chemical synthesis of a slightly modified bacterial chromosome, achieved by Craig Venter in 2010, and his demonstration that this chromosome, when reinjected into a bacterium, enabled it to survive and multiply, was only a very first step toward the synthesis of an artificial living organism. The result was more significant for the experimental possibilities it opened up than it was in itself.

These projects were inspired by a desire to naturalize life—that is, to find physical and chemical explanations for all the characteristics of living things (which had been the ambition of the first molecular biologists). Synthetic biology has the ambition to blur the boundary between living and nonliving—to the point that discussing what it means to be "living" will become futile.

This reductionist tendency has been seen at another level in the study of biological phenomena: that of human behavior and its resemblance to animal behavior. Growing interest in game theory, created by Neumann and Oskar Morgenstern (1902–1977), and applied to biology especially by Maynard Smith and Dawkins, led to the attribution of "strategies" to living organisms, and even to genes.[42] Sociobiology, made famous by the

42. Maynard Smith, 1982.

entomologist Edward Wilson (1929–) in his 1975 book *Sociobiology: The New Synthesis*, despite the criticism it received, was the harbinger of this comparison of human behaviors to those of animals governed by natural selection.[43] From the 1990s, evolutionary psychology took over, but didn't always avoid the simplistic interpretations that sociobiology became embroiled in when it attempted to explain the origin of human behaviors. Ethologists contributed to this view of a human/animal association. They showed that certain behaviors previously considered purely human traits were shared by some animal species. The 1967 book by Desmond Morris (1928–), *The Naked Ape*, and the many television series he produced popularized these theories. Let's leave aside language learning by chimpanzees, because the spectacular results obtained in the 1970s have recently been contested. But the existence of phenomena of cultural transmission in apes—such as the rapid spread of "potato washing" among macaques on the Japanese island of Koshima in the 1960s—and more recently the demonstration that analogous phenomena exist among birds, as well as animals' use of tools and their ability to calculate and make collective decisions, have been amply confirmed through field studies in natural conditions by researchers such as Jane Goodall (1934–) and Frans de Waal (1948–).

This reductionist trend, however, comes up against an opposing movement, one that highlights the diversity and complexity of the living world. Ecology reflects this movement well. After the success of the physiological ecology proposed by Lindeman and the Odum brothers, Hutchinson and Robert H. MacArthur (1930–1972)[44] developed what might be called a

43. Wilson, 2000.
44. Odenbaugh, 2013.

niche ecology, or an ecology of community, whose objective was to study the diversity of species cohabitating in what might appear to be the same ecological niche by "simple" modeling. In the same direction, in 1967 MacArthur and Wilson coauthored the book *The Theory of Island Biogeography*, whose aim was to bring together biogeography and modern ecology.[45]

MacArthur's work was at the interface between ecology and evolutionary biology, and it represented the coming together of those two disciplines. The importance attributed to "life histories,"[46] descriptions of how organisms adapt their demography to the conditions of their environment, also exemplifies this movement.

Some scientists have attempted to return to the organismic theory of ecosystems proposed by Clements. In 1974, the British chemist James Lovelock (1919–) proposed the Gaia hypothesis of a living Earth, which became very popular a few years later. More recently, he has provided a more cautious reinterpretation of it, making Gaia a complex system in which the organic and inorganic worlds are closely connected.[47] The current debate on whether an ecosystem can be described as "sick" can be seen as a belated manifestation of the organismic theory of ecosystems.

Ecological models have had their critics. Daniel Simberloff (1942–) reproached his ecologist colleagues for not submitting their models to sufficient experimentation.[48] In the 1970s, while Gould was working on evolutionary theory, Simberloff suggested that the observations made could also be explained

45. MacArthur and Wilson, 1967.

46. Korfiatis and Stamou, 1994.

47. Lovelock, 2000.

48. Dritschild, 2008.

by stochastic variations, or a "null model." Since the 1980s, these "neutral" models, which reduce the importance of competition, have found a place in ecology as they did in evolutionary biology.

Ecology has benefited from the advent of the atomic age. The study of changes to the environment caused by testing of atomic bombs and by the nuclear industry has been an important part of the work of ecologists. More recently, ecology has been confronted with growing public interest in the protection of the environment, and the success of the 1962 book by Rachel Carson (1907–1964) *Silent Spring* was one of the first signs of a new ecological awareness.[49] Though ecology has enjoyed the strong support of the general public, its models have not always enabled it to provide simple and practical solutions.

The emphasis on the diversity and complexity of the living world can also be seen at the crossroads of ecology, ethology, evolutionary biology, the cognitive sciences, and molecular biology in the many studies on communication among living beings: the characterization of the nature of exchanged signals; the meaning of the information transmitted; research into the evolutionary establishment and stabilization of these forms of communication.

Diversity is encountered at every level of the living world. By comparing the sequences of genes coding for ribosomal RNA, in 1977 Carl Woese (1928–2012) demonstrated that the tree of life consists of three branches rather than two: cells possessing a nucleus and the multicellular organisms that came from them; "traditional" bacteria, which don't have a nucleus; and archaea, single-celled organisms whose cells also lack nuclei. More recently, the direct systematic sequencing of micro-

49. Carson, 1962.

organisms present in the oceans has revealed a previously unseen diversity, identifying new microorganisms that had eluded earlier investigation as appropriate culture conditions for them had not been found. The diversity of bacterial viruses—bacteriophages—seems even greater.

The increasingly frequent use of DNA sequencing to identify and classify organisms and to establish phylogenies comes shortly after the cladistic revolution in taxonomy. The German entomologist Willi Hennig (1913–1976) proposed reconsidering the the traits traditionally used for classification to emphasize those inherited through common descendance, where individuals that carry a shared trait form a "clade."[50] The requirements of cladistics are compatible with the methods used to establish molecular phylogenies.

Human evolution has also revealed a great diversity. The sequencing of the genome of Neanderthal man and of that of his Denisovan relative has shown that these two early hominids, with already well-developed cognitive capacity and cultural behavior, were cousins who had partially exchanged their genes with *Homo sapiens*.

Finally, biological diversity is seen in the improbable alignment of plant biology and animal biology. The rise of molecular biology—with the discovery that the same fundamental biological mechanisms exist in both plants and animals—and the sequencing of the first plant genome, that of *Arabidopsis thaliana*, in 2000, had raised hopes of a greater collaboration between those two branches of biology. This has not occurred, even though analogous studies, such as research on communication signals exchanged between organisms, are conducted by both. But there are great differences between the two areas of

50. Hennig, 1966.

biology: in the signals, in the mechanisms that control embryonic development—the embryonic development of plants (or "embryonic developments," as there isn't a single model) is not homologous to that of animals. The comparison itself can be a source of confusion. The importance of epigenetic phenomena in the plant world does not imply the same is true in the animal world: the marks are different, as are their roles. The intergenerational transmission of epigenetic marks in plants is linked to their unique mechanisms of reproduction, and provides no clue to what occurs in animals.

The development of astrobiology is much more a consequence of the conquest of space and the creation of NASA than of the transformations in biology, although the founder of exobiology (the ancestor of astrobiology) was the molecular biologist Joshua Lederberg.[51] The search for life on other planets has become a major goal in astrophysics. But what forms of life should be sought? The aim of proving the existence of intelligent, "communicating" life, as in the SETI (search for extraterrestrial intelligence) program, has given way to a search for simpler organisms of bacterial type. The characteristics of these life forms are imagined using models and hypotheses offered to explain the appearance of life on Earth. The model proposed by Alexander Oparin in the 1920s and the experiments on the chemical origin of life conducted by Stanley L. Miller (1930–2007) in 1953 led in the early 1980s to the concept of the "RNA world." Will the living beings potentially found on other planets contain DNA, RNA, or other types of macromolecules? Though there is no reason to limit the search for extraterrestrial life to creatures that share our characteristics, there is still a strong temptation to look for the familiar.

51. Dick and Strick, 2005.

Historical Overview

The Dogma and Its Overturning: The Example of Prions

Since its introduction by Francis Crick in 1957,[52] the central dogma of molecular biology has been challenged many times: in 1970 with the discovery of reverse transcriptase, which generates DNA from RNA, and also with the observation of epigenetic phenomena.

The discovery of prions, pathogenic agents consisting of only protein, is another such example. Molecular biology up to then seemed to have established some order in pathogenic agents, distinguishing viruses—simple molecules of packaged genetic information—from living pathogenic organisms: bacteria and parasites.

Prions, the proteins responsible for "mad cow disease," proved to be pathogenic agents of a new type, with no genetic information, and not considered to be living organisms. Particularly resistant to all protein-denaturing agents, they catalyze a conformational change of cellular proteins similar to themselves, leading to the alteration of their function or their inactivation.

Contamination by prions is a rare phenomenon, limited to a few cases in nature such as the transmission of the disease scrapie in sheep, and to situations created by human beings—in medical or agricultural practices that have caused the spread of these pathogenic agents. While the "prion phenomenon" is no longer seen as a threat to public health, it has not left the field of biology because prions seem to have had a role in the adaptation of organisms to their environment.

52. Crick, 1958.

Molecular Noise

As we have seen, it has been only in the last twenty years or so that the study of "molecular noise" has become an important area of research, with the development of technology enabling real-time measurement of the synthesis of individual molecules, proteins, or RNA.

The noise occurs because of the small number of molecules and macromolecules present in a cell and because some reactions, such as the start of transcription, are slow, leading to the irregular production of messenger RNA molecules.

At this scale, and taking into account the speed of reactions, the occurrence of such stochastic events was predictable. However, the stochasticity of molecular phenomena was only very rarely mentioned in the writings of the first molecular biologists. Their descriptions mostly suggested that molecular processes obey a rigorous determinism. It would take experimental proof for the importance of molecular noise to be realized.

The study of molecular noise is rapidly becoming the subject of choice of many physicists who have recently converted to biology; indeed, the expertise necessary for its study and interpretation already existed in physics and engineering.

Molecular noise can be approached in three different ways. The first is to seek its origin—to reveal the mechanisms that engender it and give it its form. The second is to investigate how organisms have coped with it and the mechanisms put into place during evolution to minimize its impact. The third approach, which Gould would have called "Panglossian," seeks to determine the benefits that organisms may have derived from its existence—for example, creating a diversity of phenotypes from the same genotype, or initiating a process of cellular differentiation.

The provisional conclusions of early studies are that molecular noise exists, and that organisms have coped with it—often by limiting its negative impact, but sometimes, more rarely, by using it.

Does Systems Biology Have a Place?

The systems approach, initially proposed by Ludwig von Bertalanffy (1901–1972), is today omnipresent in all fields of biology, from molecular biology to ecology. Its importance reflects the renewed awareness of the complexity and diversity of living phenomena. It is also the outcome of the implementation of new means of observation and study, which have generated an overwhelming accumulation of data, but also of the development of computers able to deal with this mass of information.

Finally, it reflects a conviction that most of the molecular components of life and their functions have already been described, and that the work that remains is in assembling this basic knowledge into a coherent whole. But recent years have shown that this may be an illusion. The discovery of interfering RNA, and then of regulatory RNA (microRNA and, more recently, long noncoding RNA), has revealed that a group of molecules had escaped the attention of biologists,[53] as the methods routinely used in labs were not able to uncover their existence. But there were other reasons for the obliviousness of researchers. The protective phenomena induced in plants by viral infections, today interpreted as the production of interfering RNA, were described at the beginning of the twentieth century as "immunological" phenomena—without any specific mechanism being proposed, and despite the evidence that immunological

53. Morange, 2008a.

phenomena analogous to those described in animals are not present in plants.[54] As for regulatory RNA, Jacob and Monod had put forward such a hypothesis to account for the properties of the repressor, before they discovered that the latter was in fact a protein. Perhaps it was that first failure that explains why biologists ignored the hypothesis for so long, despite the antisense strategy—producing RNAs with complementary sequences to control the action of targeted genes—being widely used.

Beyond the use of the term, it isn't clear what the systems approach might bring to biology. Early results suggest that systems biology will allow biologists to go beyond the still very static view of molecular phenomena, and will reveal complex dynamics essential for understanding cellular differentiation, for example. But beyond that? Will this work give the notion of emergence a more precise meaning? Does "complexity" itself exist as a subject for study, for which models and theories will be developed, and which will shed new light on the phenomena of the living world? Or is complexity just the label we give to that which escapes us, a limited concept, or perhaps a heterogeneous group whose explanation will require a different approach each time?

An additional layer of confusion was added when systems biology was turned toward the holist, as opposed to the reductionist, approach. Indeed, the definition of a system says nothing about the approach that would best enable its study. And, in most cases, the work that is carried out is an inextricable mixture of holistic and reductionist approaches, going from bottom to top ("bottom-up") or from top to bottom ("top-down"). In the practice of research, pluralism is essential.[55]

54. Morange, 2012.
55. Mitchell, 2002.

With the rise of systems biology came the hope that mathematics would assume greater importance in the description and explanation of biological phenomena. We have seen the minor impact that the mathematical model of phyllotaxy had in the nineteenth century. By contrast, mathematical models took an important place in biology in the twentieth century, with the Hodgkin–Huxley model in physiology, but even more so in population genetics and ecology.

But some plans to formalize the phenomena of life have suffered serious failures. The model of morphogenesis developed by D'Arcy Thompson (1860–1948) inspired the dreams of biologists more than it did their experiments.[56] The mathematical biology of Nicolas Rashevsky (1899–1972) had no impact.[57] Only the mathematical model of morphogenesis proposed by Alan Turing in 1952,[58] forgotten for a long time, has been the subject of a recent renewal of interest, within the framework of evo-devo research.

Beyond Specificity?

The notion of specificity has become increasingly prevalent in the biological literature since the nineteenth century, and in the twentieth century there were many examples to illustrate it: enzymes recognizing their specific substrates and antibodies associating with their specific targets; the same was true of medicines, hormones, and neurotransmitters. Gene mutations also had "specific" effects.

In many cases, the vague notion of specificity can be substituted with the more precise term "stereospecificity": the active

56. Thompson, 2014.
57. Abraham, 2004.
58. Turing, 1952.

site of an enzyme is stereospecific because its shape is complementary to that of the substrate molecule, which allows the formation of weak bonds between the substrate and the active site of the enzyme. But this definition of stereospecificity has recently been challenged. For example, certain enzymes might be capable of recognizing several different substrates. For many biological macromolecules it would be necessary to replace the notion of stereospecificity with that of "promiscuity."

Along with the notion of specificity there was the associated idea that every component of a living organism has a precise function. The concept of promiscuity, on the other hand, means that each component has several functions. The term "tinkering," proposed by François Jacob to describe the action of evolution,[59] goes well with this new view, although multifunctionality does not necessarily mean there is an absence of specificity.

The idea that each component of a living organism might have not one but several functions, and conversely that the functions of different components could overlap, wasn't new. The pleiotropic effect of genes was described very early by Morgan and his students. Many biologists today ascribe a major role to functional redundancy in the stability of organisms and in their resistance to stressors in their environment.

The debate has been going on for a long time: it is the same one that set Praxagoras and Erasistratus on one side and Galen on the other.[60] The first two thought that the veins and arteries they had been instrumental in distinguishing had to have precise and different functions: the veins carried blood, and the arteries carried pneuma. Galen, on the other hand, proposed

59. Jacob, 1977.
60. Boylan, 2007.

that both types of vessels carried blood; the veins also transported nutritive substances provided through food, and the arteries the air provided by the lungs and also the skin.

In the face of such recurrent debates, it is difficult not to mention the *themata* described by Gerald Holton (1922–): opposing interpretations of phenomena, between which scientific theories and models seem regularly to fluctuate.[61]

Time and Life

A fully historical view of biological processes and the evolution of the living world, leaving enough room for contingency, has not yet materialized. But one could emerge and rapidly assume an important place in two realms from which, until now, history has been absent: the study of the first stages of life, and the characterization of certain diseases. Since Oparin and Haldane, it has been recognized that the "path to life" was a succession of well-defined stages: the formation of the first building blocks of life, their assembly into macromolecules, and the appearance of autonomous cells capable of reproducing. The gradual discovery of the importance of RNA, and the hypothesis proposed in the mid-1980s of a living RNA world that preceded the current living world of DNA and proteins, has made it easier to believe in that succession of stages: it has resolved the difficulty represented by the simultaneous formation of two different macromolecules, DNA and proteins, and of the precise relationship between them. Once the physicochemical conditions had been met, the formation of living beings was inevitable. This assertion supports current research in astrobiology, which aims to prove the presence of life on planets other than Earth.

61. Holton, 1978.

But did it really happen that way? Perhaps the appearance of life was no more inevitable, no more inscribed in the physicochemical conditions that reigned on the Earth's surface, than was the appearance of this or that particular animal species. Perhaps everything was the result of contingent events whose probability wasn't zero, but was indeterminable. And perhaps the research of astrobiologists is doomed to failure.

This historical view might also transform the way we look at many diseases. The progression of a cancer or a neurodegenerative disease is often seen as a predetermined and unavoidable process, which only the active intervention of a doctor can stop, slow down, or modify. But this view corresponds neither to the experience of patients, nor to the observations of doctors, nor even to the current knowledge of pathological mechanisms. If the development of a cancerous tumor is an open evolutionary process, then the particular characteristics of that tumor—and the means to halt its growth—are not decided with precision until the tumor has grown.

Mastering the Evolutionary Future

The above example of diseases shows that an evolutionary process, being a history, is open-ended. This doesn't mean, however, that it is completely unpredictable, or that it cannot be mastered. Two examples will show that the hope of biologists is not to describe the future, but to define the space that it will occupy.

The existence of the bird flu virus (and, needless to say, many other viruses) is a threat for human populations. In the past, such viruses have been able to adapt to a new host, the human being, and to cause terrible pandemics, such as the Spanish flu of 1918–1919. Recent work has attempted, with at least partial success, to define the nature of the mutations that would enable such adaptation. We will leave aside the polemics on the dangers that an experimentally modified flu virus might present to the researchers who manipulate it, and to the entire population in the event the virus escapes the laboratories where it is being studied. What is important is that such work reveals the possibility of anticipating the evolutionary paths those viruses might take, although without predicting the precise path, or paths, that will be followed, nor the probability that they will be followed in a given amount of time.

While the case of viruses is favorable, given the small size of their genome, this hope of being able to outline the possible evolutionary possibilities for a given species seems to fit with the description of gene regulatory networks described by Eric H. Davidson (1937–2015). His objective was not to trace an evolutionary path, but, taking into account the structural and functional constraints of genomes, to sketch the evolutionary landscape accessible to the organisms under study—while noting that such evolution never loses its contingent nature.

The Mystery of Life

At the beginning of his 1970 book *Chance and Necessity*, Monod asserted that the secret of life had, in large part, been unveiled. And the goals of synthetic biology showed that many researchers in that domain shared his conviction.

This didn't mean that the phenomenon of "life" had disappeared or would disappear in the near future. To naturalize life is not to make the specific characteristics of living things disappear, but to be able to explain their nature and genesis in physicochemical terms. In recent years, there has been a shift from the first type of question (on nature) to the second (on genesis). The question that is now posed, and which is far from having been answered, is that of the nature of the successive events that led to the appearance of the first living cell, whose properties today are relatively well defined.

The Ever-Ambiguous Place of the Human Being

We have described the many results that have accumulated over the past few decades that have called into question the "uniqueness" of humans: the small genetic distance from our closest cousins, the chimpanzees, and the discovery in many animal species of behaviors until recently considered to be "typically" human.

At the same time, and completely independently, the study of the disturbances to the Earth's systems caused by human activity has led many scientists to propose that a new era be added to the list of geologic eras: the Anthropocene, so named because the changes to the biosphere that have occurred during its designated span of time have been primarily the result of human activity. Poor human being, tossed between insignificance and responsibility—if not guilt—in the face of events of such major importance!

Contemporary Relevance

Today, research in biology is so different from what it was just a few years ago that attempting to find in the past the roots of present understandings may seem hopeless.

Modern research is different in the way the work is conducted. Current approaches often consist in accumulating data, from which information is then extracted, in particular through comparisons with earlier data, rather than in forming hypotheses that one then seeks patiently to confirm. The very nature of the work is different. Instead of spending long hours in the lab, biologists now use those hours to consult increasingly large databases on their computers. From being "wet," experiments are now "dry." Furthermore, research is increasingly performed in collaboration, by large multidisciplinary teams, and its results reported in publications in which lists of authors have become increasingly long.

Finally, the research carried out today most often corresponds to targeted projects with well-defined applications, supported by funding from multiple sources. The funding is often explicitly reserved for work in networks of labs spread over several countries, which also contributes to the ever-increasing number of authors for any published results.

These observations are real. But has the connection between biological research and its past been broken? No, because some of the characteristics of the research described above have existed for some time. Let's recall some of the episodes in the history of biology described above. The samples of unknown animal and plant species collected in newly explored territories during the seventeenth century, and the attempts to classify them, engendered a form of science "driven by data more than by hypotheses," to use a current turn of phrase. Discoveries were also rapidly applied, or "translated," for better or worse, as seen in the unfortunate attempts at blood transfusion in the same century. And Darwin's network of correspondents does not pale in comparison with current networks of laboratories. The image we have today of the sciences is also partially distorted, since the media and other commentators on science quite naturally focus only on what seems new.

Have science and scientists forgotten their past? In appearance, perhaps; but this is often an illusion deriving from the apparent ignorance of those involved, who have a tacitly shared stake in results. For example, the importance assigned to epigenetics today can be understood only when it is placed in the context of the long history of biology, when it is viewed alongside the rise of genetics and the great opposition once elicited by that rise.

My goal has not been to assert that science is trapped in its past. It has been simply to recall that the transformations of science are historical processes. This does not create a "burden of the past," but a complex dynamic whose description is essential for an understanding of the sciences—and of biology in particular—today.

In Conclusion

SETTING OUT TO WRITE a complete history of biology, one that includes every branch, is more than just an arduous task. For such a history can reveal the complex dynamics of the development of biological knowledge.

These dynamics include rapid innovations, as can be seen in the way biologists have repeatedly borrowed imagery from the machines that surrounded them. Harvey compared the heart to a pump, Descartes the bodies of animals to a church organ, Sherrington compared nerves to telegraph wires, and George Beadle likened metabolic pathways to assembly lines in factories.[1] More recently, the information contained in genes has been compared to the data stored in a computer, and metabolic pathways and networks, as well as those of cell signaling, to the internet and social networks.

Similarly, a rapid succession of convergent observations often serves as evidence for the establishment of new models. Models can also be the result of processes of coalescence, of the bringing together, to explain a phenomenon, of schemas that have recently been developed by different disciplines. This is how George Beadle and Boris Ephrussi attempted to establish

1. Beadle, 1948.

(without success) a relationship between genes, hormones, and embryonic induction.

In addition to these opportunities that were offered to scientists, and which they quickly exploited, is the interdisciplinary transfer of methodology, which has recurred often throughout the history of biology. Galileo and Descartes, but also the physiologists of the nineteenth century, "developmental mechanics" at the end of the same century, and geneticists and molecular biologists in the twentieth century—all borrowed methods from physics in order to further biological knowledge. Similarly, the mechanistic model of life, the comparison of an organism to a machine, persisted from Aristotle and Galen to molecular biology, via the mechanists of the seventeenth century. Explaining the characteristics of organisms through the existence of internal structures with distinct functions is also linked to the mechanistic concept of life, and appears throughout the history of biology, from antiquity to current research on functions of macromolecules.

These mechanistic ways of thinking about biological questions and of seeking answers to them can make the acceptance of new theories, such as cell theory, more challenging. The fact that cells are the basic element in the organization of living things is difficult to reconcile with the view favored by physiologists in the nineteenth century of the organism as a collection of distinct functional parts.

Some cumulative approaches have also had a long life. At the beginning of the nineteenth century, the objective of many organic chemists was to gradually describe all the chemical components of life; they were convinced that such a description would in itself provide an explanation of the phenomena unique to living organisms. Crystallographers in the twentieth century, such as Rosalind Franklin and Max Perutz, pursued the same project.

Such continuities and recurrences have sometimes appeared in pairs, as Gerald Holton has demonstrated in physics.[2] In biology, mechanistic explanations have always been contrasted with explanations by the activity of fermentation. Such explanations were used by Aristotle to account for embryonic development; they were used (and abused) by alchemists; the enzymatic theory of life flourished in the first half of the twentieth century. We believe that modern explanations that call on self-organization and epigenetic modification have a great deal in common with these earlier explanations.

Along the same lines, the opposition that Ernst Mayr clearly established at the beginning of the 1960s between functional biology and evolutionary biology, between the search for proximate causes and that for ultimate causes, is very old.[3] It was already clearly visible in Haeckel and Darwin, and it retains an important place in contemporary biology.

To be able to fully present the historical dynamics of the development of biological knowledge, we must add at least four other elements. The first is the existence of points of reference, often called "icons," which sociologists of science have also named "totems": apparatuses, experiments, models, or theories that assume an enduring and indisputable place in biology. For example, in the eighteenth century, Newton's theory became a required reference for biologists and chemists, as it was for physicists.

In contrast, sometimes contingent events have, for years or even decades, influenced the research and explanatory models of biologists. Hans Driesch, through his role in the birth of

2. Holton, 1978.

3. Mayr, 1961.

developmental mechanics and the scientific reputation he earned from it, was able to encourage the development of a holistic approach in biology, particularly in embryology. Similarly, the development of high-resolution genetic mapping, which accompanied the early sequencing of the human genome, at the end of the 1980s enabled the isolation of genes whose modifications were responsible for certain human illnesses or behavioral anomalies. This saved the idea of "the gene for," which should have disappeared with the progress achieved in the description of the structure of genes and their mechanisms of activity.

Even more important was the complex interaction between contemporary culture and scientific knowledge, and the circulation of that knowledge between different disciplines, a circulation that was more viscous than fluid. For example, a model might be adopted by one discipline at the very moment it is discarded by the discipline that developed it. The thermodynamics approach has had an equally strong influence on both biochemistry and ecology. But ecologists adopted the approach in the 1940s, while biochemists had already criticized its reductionist concept of life—reducing its development and its requirements to a single parameter—in the 1910s. Conversely, molecular biology is not a result of quantum physics, except through the progress in chemistry that quantum physics brought about. The physics that played a part in the development of molecular biology—for the creation of complex machines such as ultracentrifuges—was "classic" physics. Nonetheless, the prestige acquired by quantum physics inspired many young scientists to try to use it to breathe new life into other scientific disciplines, and they hoped especially to make important contributions to the field of biology.

It is not just the circulation of knowledge that has been viscous; the development of scientific work can be described in the same way. A line of research might be stubbornly pursued not because it enables new questions to be answered, but simply because it has momentum, justifying the data it produces only by their accumulation.

Finally, we mustn't forget that there have also been true breakthroughs in biology, analogous to Thomas Kuhn's scientific revolutions:[4] the discovery that there was a history of life transformed biology irreversibly.

From all of this there results a complex dynamic, for which Fernand Braudel, in the field of general history, gave only a somewhat oversimplified image.[5] It is this complex dynamic that is the basis of the autonomous development of biological knowledge. "Autonomy" here doesn't mean that these scientific developments were independent of the contexts in which they occurred, but rather that a context is never enough to explain the genesis of such a development. Schelling's *Naturphilosophie* prepared the way for the acceptance of cell theory. But the invention of the achromatic microscope and the discovery of the nucleus and the nucleolus contributed just as much, if not more. Similarly, the relationship between animal biology and plant biology—sometimes close and sometimes nonexistent, complex in its detail, with exchanges that may be rapid or slow—is the result of dynamics that are sometimes shared by both and sometimes unique to each of these branches of biology. In this, biology is no different from the other sciences, but the diversity of its objects of study, and of its subdisciplines

4. Kuhn, 2012.

5. Braudel, 1992.

that study them, makes the dynamics in biology particularly complex.

The oppositions that philosophers have noted between, for example, the reductionist and holistic approaches to phenomena of life do exist, but they are caught up in this complex dynamic of the history of biology, and are in part distorted by it. The history of biology is indeed a history, with all the contingency that the term implies.

REFERENCES

Abraham, T. H. (2004) "Nicolas Rashevsky's mathematical biophysics," *J. Hist. Biol.*, 37, pp. 333–385.

Ackert, L. T., Jr. (2006) "The role of microbes in agriculture: Sergei Vinogradskii's discovery and investigation of chemosynthesis, 1880–1910," *J. Hist. Biol.*, 39, pp. 373–406.

Ackert, L. T., Jr. (2007) "The 'cycle of life' in ecology: Sergei Vinogradskii's soil microbiology, 1885–1940," *J. Hist. Biol.*, 40, pp. 109–145.

Acot, P. (1998) *The European Origins of Scientific Ecology*, Philadelphia, Gordon and Breach.

Adler, I., Barabe, D., and Jean, R. V. (1997) "A history of the study of phyllotaxis," *Ann. Bot.*, 80, pp. 231–244.

Allen, G. E. (1969) "Hugo de Vries and the reception of the 'mutation theory,'" *J. Hist. Biol.*, 2, pp. 55–87.

Allen, G. E. (1975) *Life Sciences in the Twentieth Century*, Cambridge, Cambridge University Press.

Allen, G. E. (1978) *Thomas Hunt Morgan: The Man and His Science*, Princeton, NJ, Princeton University Press.

Allen, G. E. (2004) "A pact with the embryo: Viktor Hamburger, holistic and mechanistic philosophy in the development of neuroembryology, 1927–1955," *J. Hist. Biol.*, 37, pp. 421–475.

Allen, G. E. (2008) "Rebel with two causes: Hans Driesch," in O. Harman and M. R. Dietrich (eds.), *Rebels, Mavericks, and Heretics in Biology*, New Haven, CT, Yale University Press, pp. 37–64.

Allin, M. (1998) *Zarafa: A Giraffe's True Story from Deep in Africa to the Heart of Paris*, London, Walker.

Anderson, L. (1976) "Bonnet's taxonomy and the chain of being," *J. Hist. Ideas*, 37, pp. 45–58.

Appel, T. A. (2000) *Shaping Biology: The National Science Foundation and American Biological Research, 1945–1975*, Baltimore, MD, Johns Hopkins University Press.

Aristotle (1955) *Parva naturalia*, ed. David Ross, Oxford, Clarendon.
Aristotle (1982) *De generatione et corruptione*, trans. C.J.F. Williams, Oxford, Clarendon.
Aristotle (1991) *History of Animals*, ed. and trans. D. M. Balme, Cambridge, MA, Harvard University Press.
Barthélémy-Madaule, M. (1979) *Lamarck ou Le mythe du précurseur*, Paris, Éditions du Seuil.
Bashford, A., and Levine, P. (eds.) (2012) *The Oxford Handbook of the History of Eugenics*, Oxford, Oxford University Press.
Bateson, W. (1894) *Materials for the Study of Variation Treated with Especial Regard to Discontinuity in the Origins of Species*, Baltimore, MD, Johns Hopkins University Press.
Beadle, G. W. (1948) "The genes of men and molds," *Sci. Am.*, 179 (3), pp. 30–39.
Bechtel, W. (2008) *Discovering Cell Mechanisms: The Creation of Modern Cell Biology*, Cambridge, Cambridge University Press.
Bensaude-Vincent, B. (1993) *Lavoisier*, Paris, Flammarion.
Berche, P. (2007) *Une histoire des microbes*, Montrouge, John Libbey.
Bernard, C. (1974) *Lectures on the Phenomena of Life Common to Animals and Plants* [1878], trans. H. E. Hoff, R. Guillemin, and L. Guillemin, Springfield, IL, Thomas.
Bernard, C. (2018) *An Introduction to the Study of Experimental Medicine* [1865], trans. H. C. Creene, New York, Dover.
Bernier, G. (2013) *Darwin: Un pionnier de la physiologie végétale*, Bruxelles, Académie royale de Belgique.
Bertoloni Meli, D. (2011) *Mechanism, Experiment, Disease: Marcello Malpighi and Seventeenth-Century Anatomy*, Baltimore, MD, Johns Hopkins University Press.
Bichat, X. (1977) *Physiological Researches on Life and Death* [1802], trans. F. Gold, New York, Arno.
Blanckaert, C. (ed.) (2013) *La Vénus hottentote: Entre Barnum et Muséum*, Paris, Muséum national d'histoire naturelle, Publications scientifiques.
Bloch, M. (1941–1943) "Apologie pour l'histoire ou métier d'historien," in M. Bloch, *L'histoire, la guerre, la résistance*, Paris, Gallimard, pp. 843–985.
Bonneuil, C. (2006) "Mendelism, plant breeding and experimental cultures: Agriculture and the development of genetics in France," *J. Hist. Biol.*, 39, pp. 281–308.
Bonneuil, C., and Fressoz, J.-B. (2013) *L'événement anthropocène: La Terre, l'histoire et nous*, Paris, Éditions du Seuil.
Borrell, M. (1985) "Organotherapy and the emergence of reproductive endocrinology," *J. Hist. Biol.*, 18, pp. 1–30.
Borrello, M. (2008) "Dogma, heresy, and conversion: Vero Copner Wynne-Edwards's crusade and the levels-of-selection debate," in O. Harman and M. R. Dietrich

(eds.), *Rebels, Mavericks, and Heretics in Biology*, New Haven, CT, Yale University Press, pp. 213–230.

Boustani, F. (2007) *La circulation du sang: Entre Orient et Occident, l'histoire d'une découverte*, Paris, Philippe Rey.

Boutibonnes, P. (1994) *Van Leeuwenhoek: L'exercice du regard*, Paris, Belin.

Bowler, P. J. (1971) "Preformation and pre-existence in the seventeenth century: A brief analysis," *J. Hist. Biol.*, 4, pp. 221–244.

Bowler, P. J. (1977) "Edward Drinker Cope and the changing structure of evolutionary theory," *Isis*, 68, pp. 249–265.

Bowler, P. J. (1983) *Eclipse of Darwinism*, Baltimore, MD, Johns Hopkins University Press.

Boylan, M. (2007) "Galen: On blood, the pulse, and the arteries," *J. Hist. Biol.*, 40, pp. 207–230.

Brannigan, A. (1979) "The reification of Mendel," *Soc. Stud. Sci.*, 9, pp. 423–454.

Braudel, F. (1992) *The Mediterranean and the Mediterranean World in the Age of Philip II* [1949], trans. S. Reynolds, ed. and abr. R. Ollard, New York, Harper and Collins.

Braun, R. (2011) "Accessory food factors: Understanding the catalytic function," *J. Hist. Biol.*, 44, pp. 483–504.

Breidbach, O. (2006) *Visions of Nature: The Art and Science of Ernst Haeckel*, New York, Prestel.

Brigandt, I. (2005) "The instinct concept of the early Konrad Lorenz," *J. Hist. Biol.*, 38, pp. 571–608.

Brock, T. D. (1990) *The Emergence of Bacterial Genetics*, Cold Spring Harbor, NY, Cold Spring Harbor Laboratory.

Brock, T. D. (1999) *Milestones in Microbiology, 1546–1940*, Washington, DC, ASM.

Browne, J. (2010a) *Charles Darwin: Voyaging*, Princeton, NJ, Princeton University Press.

Browne, J. (2010b) *Charles Darwin: The Power of Place*, Princeton, NJ, Princeton University Press.

Buffetaut, E. (2002) *Cuvier: Le découvreur de mondes disparus*, Paris, Pour la Science.

Buffon (1986), *De la manière d'étudier et de traiter l'histoire naturelle* [1749], Paris, Bibliothèque nationale.

Burkhardt, R. W., Jr. (1977) *The Spirit of System: Lamarck and Evolutionary Biology*, Cambridge, MA, Harvard University Press.

Burkhardt, R. W., Jr. (2005) *Patterns of Behavior: Konrad Lorenz, Niko Tinbergen, and the Founding of Ethology*, Chicago, University of Chicago Press.

Cain, J. (2009) "Rethinking the synthesis period in evolutionary studies," *J. Hist. Biol.*, 42, pp. 621–648.

Cairns, J., Stent, G. S., and Watson, J. D. (1992) *Phage and the Origins of Molecular Biology*, Cold Spring Harbor, NY, Cold Spring Harbor Laboratory.
Camardi, G. (2001) "Richard Owen, morphology and evolution," *J. Hist. Biol.*, 34, pp. 481–515.
Canguilhem, G. (1965a) "Aspects du vitalisme," in *La connaissance de la vie*, Paris, Vrin, pp. 83–100.
Canguilhem, G. (1965b) "Machine et organisme," in *La connaissance de la vie*, Paris, Vrin, pp. 101–127.
Canguilhem, G. (1965c) "La théorie cellulaire," in *La connaissance de la vie*, Paris, Vrin, pp. 43–80.
Canguilhem, G. (1968) "L'homme de Vésale dans le monde de Copernic," in *Études d'histoire et de philosophie des sciences*, Paris, Vrin, pp. 27–35.
Carlson, E. A. (2001) *The Unfit: A History of a Bad Idea*, Cold Spring Harbor, NY, Cold Spring Harbor Laboratory.
Carpenter, K. J. (2000) *Beriberi, White Rice, and Vitamin B: A Disease, a Cause, and a Cure*, Berkeley, University of California Press.
Carroll, S. B. (2008) "Evo-devo and an expanding evolutionary synthesis: A genetic theory of morphological evolution," *Cell*, 134, pp. 25–36.
Carson, R. (1962) *Silent Spring*, Boston, MA, Houghton Mifflin Harcourt.
Chadarevian, S. de (2002) *Designs for Life: Molecular Biology after World War II*, Cambridge, Cambridge University Press.
Chappey, J.-L. (2009) *Des naturalistes en révolution: Les procès-verbaux de la Société d'histoire naturelle de Paris (1790–1798)*, Paris, CTHS.
Chen, W. (1996) *Comment Fleming n'a pas inventé la pénicilline*, Paris, Synthélabo.
Chiang, H. H.-S. (2009) "The laboratory technology of discrete molecular separation: The historical development of gel electrophoresis and the material epistemology of biomolecular science, 1945–1970," *J. Hist. Biol.*, 42, pp. 495–527.
Churchill, F. B. (1991) "The rise of classical descriptive embryology," in S. F. Gilbert (ed.), *A Conceptual History of Modern Embryology*, New York, Plenum, pp. 1–29.
Clarke, E., and Jacyna, L. S. (1987) *Nineteenth-Century Origins of Neuroscientific Concepts*, Berkeley, University of California Press.
Clements, F. E. (1916) *Plant Succession: An Analysis of the Development of Vegetation*, Washington, DC, Carnegie Institution.
Clements, F. E. (1936) "Nature and structure of the climax," *J. Ecol.*, 24, pp. 252–284.
Cobb, M. (2002) "Exorcizing the animal spirits: Jan Swammerdam on nerve function," *Nat. Rev. Neurosci.*, 3, pp. 395–400.
Cobb, M. (2006) *Generation: Seventeenth-Century Scientists who Unraveled the Secrets of Sex, Life, and Growth*, New York, Bloomsbury.

Cock, A. G., and Forsdyke, D. R. (2008) *Treasure Your Exceptions: The Science and Life of William Bateson*, New York, Springer.

Cohen, C. (1994) *Le destin du mammouth*, Paris, Éditions du Seuil.

Coleman, W. (1984) *Georges Cuvier, Zoologist*, Cambridge, MA, Harvard University Press.

Collins, D. (2009) "Misadventures in the Burgess shale," *Nature*, 460, pp. 952–953.

Comfort, N. (2001) *The Tangled Field: Barbara McClintock's Search for Patterns of Genetic Control*, Cambridge, MA, Harvard University Press.

Conry, Y. (1974) *L'introduction du darwinisme en France au XIXe siècle*, Paris, Vrin.

Corsi, P. (1988) *The Age of Lamarck: Evolutionary Theories in France, 1790–1830*, Berkeley, University of California Press.

Corsi, P. (2005) "Before Darwin: Transformist concepts in European natural history," *J. Hist. Biol.*, 38, pp. 67–83.

Corsi, P., Gayon, J., Gohau, G., and Tirard, S. (2006) *Lamarck, philosophe de la nature*, Paris, Presses universitaires de France.

Craver, C. F. (2003) "The making of a memory mechanism," *J. Hist. Biol.*, 36, pp. 153–195.

Craver, C. F., and Darden, L. (2013) *In Search of Mechanisms: Discoveries across the Life Sciences*, Chicago, University of Chicago Press.

Creager, A.N.H. (2002) *The Life of a Virus: Tobacco Mosaic Virus as an Experimental Model, 1930–1965*, Chicago, University of Chicago Press.

Creager, A.N.H. (2013) *Life Atomic: A History of Radioisotopes in Science and Medicine*, Chicago, University of Chicago Press.

Crick, F.H.C. (1958) "On protein synthesis," *Symp. Soc. Exp. Biol.*, 12, pp. 138–163.

Crick, F.H.C., and Lawrence, P. A. (1975) "Compartments and polyclones in insect development," *Science*, 189, pp. 340–347.

Cuvier, G. (1985) *Discours sur les révolutions de la surface du globe* [1825], Paris, Christian Bourgois.

Dagognet, F. (1994) *Pasteur sans la légende*, Paris, Synthélabo.

Darden, L. (1977) "William Bateson and the promise of Mendelism," *J. Hist. Biol.*, 10, pp. 87–106.

Darnton, R. (1986) *Mesmerism and the End of the Enlightenment in France*, Cambridge, MA, Harvard University Press.

Darwin, C. (1981) *The Descent of Man, and Selection in Relation to Sex* [1871], Princeton, NJ, Princeton University Press.

Darwin, C. (1988) *The Movements and Habits of Climbing Plants* [1880], New York, New York University Press.

Darwin, C. (1989) *The Expression of the Emotions in Man and Animals*, New York, New York University Press.

Darwin, C. (1998) *The Variation of Animals and Plants under Domestication* [1868], Baltimore, MD, Johns Hopkins University Press.

Darwin, C. (2001) *The Voyage of the Beagle: Journal of Researches into the Natural History and Geology of the Countries Visited during the Voyage of H.M.S. Beagle Round the World* [1839], New York, Modern Library.

Darwin, C. (2009) *On the Origin of Species* [1859], ed. Jim Endersby, Cambridge, Cambridge University Press.

Davidson, E. H. (2006) *The Regulatory Genome: Gene Regulatory Networks in Development and Evolution*, Burlington, IN, Academic Press.

Da Vinci, L. (2000) *Notebooks*, New York, Arcade.

Dawkins, R. (1976) *The Selfish Gene*, Oxford, Oxford University Press.

Deichmann, U. (2007) "'Molecular' versus 'colloidal:' Controversies in biology and biochemistry, 1900–1940," *Bull. Hist. Chem.*, 32, pp. 105–118.

Deichmann, U., Schuster, S., Mazat, J.-P., and Cornish-Bowden, A. (2013) "Commemorating the 1913 Michaelis-Menten paper *Die Kinetik der Invertinwirkung*: Three perspectives," *FEBS J.*, 281, pp. 435–463.

Deléage, J.-P. (1991) *Une histoire de l'écologie*, Paris, La Découverte.

Delisle, R. G. (2009) *Les philosophies du néodarwinisme: Conceptions divergentes sur l'homme et le sens de l'évolution*, Paris, Presses universitaires de France.

Delord, J. (2009) "Écologie et évolution: Vers une articulation multi-hiérarchisée," in T. Heams, P. Huneman, G. Lecointre, and M. Silberstein (eds.), *Les mondes darwiniens*, Paris, Syllepse, pp. 607–628.

Denis, G. (2011) "The optical Galilean interpretation of the antique Theophrastian model for plant diseases," *Galileana*, 8, pp. 183–204.

Descartes, R. (2000) *Description du corps humain*, vol. 11 of *Oeuvres complètes*, ed. C. Adam and P. Tannery, Paris, Librairie philosophique J. Vrin.

Descartes, R. (2003) *Treatise of Man* [1630–1633], trans. T. S. Hall, New York, Prometheus Books.

Dick, S. J., and Strick, J. E. (2005) *The Living Universe: NASA and the Development of Astrobiology*, New Brunswick, NJ, Rutgers University Press.

Dietrich, M. R. (1994) "The origins of the neutral theory of molecular evolution," *J. Hist. Biol.*, 27, pp. 21–59.

Djebbar, A. (2001) *Une histoire de la science arabe*, Paris, Éditions du Seuil.

Dobzhansky, T. (1937) *Genetics and the Origin of Species*, New York, Columbia University Press.

Dritschild, W. (2008) "Bringing statistical methods to community and evolutionary ecology: Daniel S. Simberloff," in O. Harman and M. R. Dietrich (eds.), *Rebels, Mavericks, and Heretics in Biology*, New Haven, CT, Yale University Press, pp. 356–371.

Dröscher, A. (1998) "1998: The centenary of the discovery of the Golgi apparatus," *Glycoconj. J.*, 15, pp. 733–736.

Drouin, J.-M. (1991) *Réinventer la nature, l'écologie et son histoire*, Paris, Desclée de Brouwer.

Duchesneau, F. (2000) *Genèse de la théorie cellulaire*, Paris, Vrin.

Duchesneau, F., Kupiec, J.-J., and Morange, M. (2013) *Claude Bernard: La méthode de la physiologie*, Paris, Éditions Rue d'Ulm.

Dugatkin, L. A. (2011) *The Prince of Evolution: Peter Kropotkin's Adventures in Science and Politics*, CreateSpace Independent Publishing Platform.

Dupont, J.-C. (1999) *Histoire de la neurotransmission*, Paris, Presses universitaires de France.

Dupont, J.-C., and Schmitt, S. (2004) *Du feuillet au gène: Une histoire de l'embryologie moderne, fin xviiie–xxe siècle*, Paris, Éditions Rue d'Ulm.

Egerton, F. N. (1983) "The history of ecology: Achievements and opportunities. Part one," *J. Hist. Biol.*, 16, pp. 259–310.

Endersby, J. (2008) *Imperial Nature: Joseph Hooker and the Practices of Victorian Science*, Chicago, University of Chicago Press.

Enenkel, K.A.E., and Smith, P. J. (eds.) (2007) *Early Modern Zoology: The Construction of Animals in Science, Literature and the Visual Arts*, Leiden, Brill.

Erwin, D. H., and Davidson, E. H. (2009) "The evolution of hierarchical gene regulatory networks," *Nat. Rev. Genet.*, 10, pp. 141–148.

Farley, J. (1974) "The initial reactions of French biologists to Darwin's *Origin of Species*," *J. Hist. Biol.*, 7, pp. 275–300.

Farley, J. (1977) *The Spontaneous Generation Controversy from Descartes to Oparin*, Baltimore, MD, Johns Hopkins University Press.

Finger, S. (1994) *Origins of Neuroscience: A History of Explorations into Brain Functions*, Oxford, Oxford University Press.

Fischer, J.-L. (1991) "Laurent Chabry and the beginnings of experimental embryology in France," in S. Gilbert (ed.), *A Conceptual History of Modern Embryology*, New York, Plenum, pp. 31–41.

Fisher, R. A. (1918) "XV: The correlation between relatives on the supposition of Mendelian inheritance," *Trans. R. Soc. Edinb.*, 52, pp. 399–433.

Fishman, M. (2004) *An Elusive Victorian: The Evolution of Alfred Russel Wallace*, Chicago, University of Chicago Press.

Forest, D. (2014) *Neuroscepticisme*, Paris, Ithaque.

Foucault, M. (1994) *The Order of Things: An Archaeology of the Human Sciences*, New York, Vintage.

Franklin, A., Edwards, A.W.F., Fairbanks, D. J., Hartl, D. L., and Seidenfeld, T. (2008) *Ending the Mendel-Fisher Controversy*, Pittsburgh, PA, University of Pittsburgh Press.

Freedberg, D. (2002) *The Eye of the Lynx: Galileo, His Friends, and the Beginnings of Modern Natural History*, Chicago, University of Chicago Press.

French, R. (1999) *Dissection and Vivisection in the European Renaissance*, Aldershot, UK, Ashgate.

Frisch, K. von (1953) *The Dancing Bees: An Account of the Life and Senses of the Honey Bee* [1927], New York, Harvest Books.

Fruton, J. S. (1999) *Proteins, Enzymes, Genes: The Interplay of Chemistry and Biology*, New Haven, CT, Yale University Press.

Garcia-Sancho, M. (2012) *Biology, Computing, and the History of Molecular Sequencing: From Proteins to DNA (1945–2000)*, Basingstoke, UK, Palgrave Macmillan.

Gaudillière, J.-P. (2005) "Better prepared than synthesized: Adolf Butenandt, Schering Ag and the transformation of sex steroids into drugs (1930–1946)," *Stud. Hist. Philos. Biol. Biomed. Sci.*, 36, pp. 612–644.

Gaukroger, S. (2010) *The Collapse of Mechanism and the Rise of Sensibility: Science and the Shaping of Modernity*, Oxford, Oxford University Press.

Gayon, J., and Burian, R. M. (2004) "National traditions and the emergence of genetics: The French example," *Nat. Genet.*, 5, pp. 150–156.

Gayon, J., and Zallen, D. (1998) "The role of the Vilmorin Company in the promotion and diffusion of the experimental science of heredity in France, 1840–1920," *J. Hist Biol.*, 31, pp. 241–262.

Gehring, W. J., and Ikeo, K. (1999) "Pax6 mastering eye morphogenesis and eye evolution," *Trends Genet.*, 15, pp. 371–377.

Gilbert, S. F. (1991) *A Conceptual History of Modern Embryology*, vol. 7 of *Developmental Biology: A Comprehensive Synthesis*, New York, Plenum.

Gillham, N. W. (2001) *A Life of Sir Francis Galton: From African Exploration to the Birth of Eugenics*, Oxford, Oxford University Press.

Goldschmidt, R. (1982) *The Material Basis of Evolution* [1940], New Haven, CT, Yale University Press.

Gould, S. J. (1977) *Ontogeny and Phylogeny*, Cambridge, MA, Harvard University Press.

Gould, S. J., and Eldredge, N. (1977) "Punctuated equilibria: The tempo and mode of evolution reconsidered," *Paleobiology*, 3, pp. 115–151.

Gould, S. J., and Lewontin, R. (1979) "The spandrels of San Marco and the Panglossian paradigm: A critique of the adaptationist programme," *Proc. R. Soc. Lond. B*, 205, pp. 581–598.

Gould, S. J., and Vrba, E. S. (1982) "Exaptation: A missing term in the science of form," *Paleobiology*, 8, pp. 4–15.

Gouz, S. (2006) "Biographie d'une vision du monde: Les relations entre science, philosophie et politique dans la conception marxiste de J. B. S. Haldane," thesis, Université Claude Bernard-Lyon 1.

Govindjee, B. T., Gest, H., and Allen, J. F. (eds.) (2005) *Discoveries in Photosynthesis*, Dordrecht, Springer.

Grmek, M. D. (1973) *Raisonnement expérimental et recherches toxicologiques chez Claude Bernard*, Geneva, Droz.

Grmek, M. D. (1990) *La première révolution biologique*, Paris, Payot.

Grmek, M. D. (1997) *Le chaudron de Médée: L'expérimentation sur le vivant dans l'Antiquité*, Paris, Synthélabo.

Guerrini, A. (2013) "Experiments, causation, and the uses of vivisection in the first half of the seventeenth century," *J. Hist. Biol.*, 46, pp. 227–254.

Hagen, J. B. (1992) *An Entangled Bank: The Origins of Ecosystem Ecology*, New Brunswick, NJ, Rutgers University Press.

Hager, T. (1995) *Force of Nature: The Life of Linus Pauling*, New York, Simon and Schuster.

Harper, P. S. (2005) "William Bateson, human genetics and medicine," *Hum. Genet.*, 118, pp. 141–151.

Harris, H. (1999) *The Birth of the Cell*, New Haven, CT, Yale University Press.

Hartl, D. L., and Fairbanks, D. J. (2007) "Mud sticks: On the alleged falsification of Mendel's data," *Genetics*, 175, pp. 975–979.

Hartmann, T. (2008) "The lost origin of chemical ecology in the late nineteenth century," *Proc. Natl. Acad. Sci. USA*, 105, pp. 4541–4546.

Heams, T., Huneman, P., Lecointre, G., and Silberstein, M. (eds.) (2009) *Les mondes darwiniens: L'évolution de l'évolution*, Paris, Syllepse.

Hennig, W. (1966) *Phylogenetic Systematics* [1950], Urbana, University of Illinois Press.

Hippocrates (2012) *On the Art of Medicine*, Leiden, Brill.

Holmes, F. L. (1974) *Claude Bernard and Animal Chemistry*, Cambridge, MA, Harvard University Press.

Holmes, F. L. (1991) *The Formation of a Scientific Life, 1900–1933*, vol. 1 of *Hans Krebs*, Oxford, Oxford University Press.

Holmes, F. L. (1993) *Architect of Intermediary Metabolism, 1933–1937*, vol. 2 of *Hans Krebs*, Oxford, Oxford University Press.

Holmes, F. L. (2001) *Meselson, Stahl, and the Replication of DNA*, New Haven, CT, Yale University Press.

Holton, G. (1978) *The Scientific Imagination: Case Studies*, Cambridge, Cambridge University Press.

Hopwood, N. (2006) "Pictures of evolution and charges of fraud: Ernst Haeckel's embryological illustrations," *Isis*, 97, pp. 260–301.

Hoquet, T. (ed.) (2005) *Les fondements de la botanique: Linné et la classification des plantes*, Paris, Vuibert.

Hoquet, T. (2009) *Darwin contre Darwin: Comment lire "L'origine des espèces"?*, Paris, Éditions du Seuil.

Horder, T. (2008) "Roger Sperry and integrative action in the nervous system," in O. Harman and M. R. Dietrich (eds.), *Rebels, Mavericks, and Heretics in Biology*, New Haven, CT, Yale University Press, pp. 174–193.

Horder, T. J., and Weindling, P. J. (1986) "Hans Spemann and the organizer," in T. J. Horder, J. A. Witkowski, and C. C. Wylie (eds.), *A History of Embryology*, Cambridge, Cambridge University Press, pp. 183–242.

Horder, T. J., Witkowski, T. A., and Wylie, C. C. (eds.) (1986) *A History of Embryology*, Cambridge, Cambridge University Press.

Howard, J.A.K. (2003) "Dorothy Hodgkin and her contributions to biochemistry," *Nat. Rev. Mol. Cell Biol.*, 4, pp. 891–896.

Hughes, S. S. (1977) *The Virus: A History of the Concept*, New York, Science History.

Hughes, S. S. (2011) *Genentech: The Beginnings of Biotech*, Chicago, University of Chicago Press.

Humboldt, A. von, and Bonpland, A. (1895) *Personal Narrative of Travels to the Equinoctial Regions of America during the Years 1799–1804*, trans. and ed. T. Ross, vols. 2 and 3, London, George Routledge, www.biodiversitylibrary.org.

Ihde, A. J., and Becker, S. L. (1971) "Conflict of concepts in early vitamin studies," *J. Hist. Biol.*, 4, pp. 1–33.

Irwin, L. N. (2007) *Scotophobin: Darkness at the Dawn of the Search for Memory Molecules*, Lanham, Hamilton Books.

Jablonka, E., and Lamb, M. J. (2005) *Evolution in Four Dimensions: Genetic, Epigenetic, Behavioral and Symbolic Variation in the History of Life*, Cambridge, MA, MIT Press.

Jacob, F. (1977) "Evolution and tinkering," *Science*, 196, pp. 1161–1166.

Jacob, F. (1988) *The Statue Within: An Autobiography*, trans. F. Philip, New York, Basic Books.

Jacob, F. (1993) *The Logic of Life: A History of Heredity* [1970], trans. B. E. Spillmann, Princeton, NJ, Princeton University Press.

Jacob, F. (2000) "Conclusions," *Bull. Acad. Natl. Méd.*, 184 (6), pp. 1237–1240.

Joly, B. (2013) *Histoire de l'alchimie*, Paris, Vuibert.

Jordan, J. M. (2016) "'Ancient episteme' and the nature of fossils: A correction of a modern scholarly error," *Hist. Philos. Life Sci.* 38, pp. 90–116.

Judson, H. F. (1996) *The Eighth Day of Creation: The Makers of the Revolution in Biology* [1979], Cold Spring Harbor, NY, Cold Spring Harbor Laboratory.

Kalikow, T. J. (1983) "Konrad Lorentz's ecological theory: Explanation and ideology," *J. Hist. Biol.*, 16, pp. 39–73.

Kant, I. (2007) *Critique of Judgment* [1790], trans. J. C. Meredith, rev. and ed. N. Walker, Oxford, Oxford University Press.

Kay, L. E. (1993) *The Molecular Vision of Life: Caltech, the Rockefeller Foundation, and the Rise of the New Biology*, Oxford, Oxford University Press.

Kay, L. E. (2000) *Who Wrote the Book of Life? A History of the Genetic Code*, Palo Alto, CA, Stanford University Press.

Keller, E. F. (2002) *Making Sense of Life: Explaining Biological Development with Models, Metaphors, and Machines*, Cambridge, MA, Harvard University Press.

Kendler, K. S., and Schaffner K. F. (2011) "The dopamine hypothesis of schizophrenia: An historical and philosophical analysis," *Philos. Psychiatry Psychol.*, 18, pp. 41–63.

Kevles, D. J. (1998) *In the Name of Eugenics: Genetics and the Uses of Human Heredity*, Cambridge, MA, Harvard University Press.

King, M.-C., and Wilson, A. C. (1975) "Evolution at two levels in humans and chimpanzees," *Science*, 188, pp. 107–116.

Kingsland, S. E. (2005) *The Evolution of American Ecology 1890–2000*, Baltimore, MD, Johns Hopkins University Press.

Kocandrle, R., and Kleisner, K. (2013) "Evolution born of moisture: Analogies and parallels between Anaximander's ideas on origin of life and man and later pre-Darwinian and Darwinian evolutionary concepts," *J. Hist. Biol.*, 46, pp. 103–124.

Koehler, P. J., Finger, S., and Piccolino, M. (2009) "The 'eels' of South America: Mid-eighteenth-century Dutch contributions to the theory of animal electricity," *J. Hist. Biol.*, 42, pp. 715–763.

Koerner, L. (1999) *Linnaeus: Nature and Nation*, Cambridge, MA, Harvard University Press.

Kohler, R. E. (1975) "The history of biochemistry: A survey," *J. Hist. Biol.*, 8, pp. 275–318.

Kohler, R. E. (1994) *Lords of the Fly: Drosophila Genetics and the Experimental Life*, Chicago, University of Chicago Press.

Kohler, R. E. (2008) *From Medical Chemistry to Biochemistry: The Making of a Biomedical Discipline*, Cambridge, Cambridge University Press.

Korfiatis, K., and Stamou, G. (1994) "Emergence of new fields in ecology: The case of life history studies," *Hist. Philos. Life Sci.*, 16, pp. 97–116.

Kuhn, T. S. (2012) *The Structure of Scientific Revolutions* [1962], Chicago, University of Chicago Press.

Lachenal, G. (2014) *Le médicament qui devait sauver l'Afrique: Un scandale pharmaceutique aux colonies*, Paris, La Découverte.

Lagier, R. (2004) *Les races humaines selon Kant*, Paris, Presses universitaires de France.

Lagueux, O. (2003) "Geoffroy's giraffe: The hagiography of a charismatic mammal," *J. Hist. Biol.*, 36, pp. 225–247.

Lamarck, J.-B. Monet de (1984) *Zoological Philosophy* [1809], trans. H. Elliot, Chicago, University of Chicago Press.

Landecker, H. (2007) *Culturing Life: How Cells Became Technologies*, Cambridge, MA, Harvard University Press.

Lantéri-Laura, G. (1970) *Histoire de la phrénologie: L'homme et son cerveau selon F. J Gall*, Paris, Presses universitaires de France.

Laporte, L. F. (1994) "Simpson on species," *J. Hist. Biol.*, 27, pp. 141–159.

Laubichler, M. D., and Davidson, E. H. (2008) "Boveri's long experiment: Sea urchin merogones and the establishment of the role of nuclear chromosomes in development," *Dev. Biol.*, 314, pp. 1–11.

Laubichler, M. D., and Maienschein, J. (eds.) (2007) *From Embryology to Evo-Devo: A History of Developmental Evolution*, Cambridge, MA, MIT Press.

Lax, E. (2004) *The Mould in Dr. Florey's Coat: The Remarkable True Story of the Penicillin Miracle*, London, Little, Brown.

Le Goff, J. (2014) *Faut-il vraiment découper l'histoire en tranches?* Paris, Éditions du Seuil.

Lennox, J. G. (2001) *Aristotle's Philosophy of Biology: Studies in the Origins of Life Science*, Cambridge, Cambridge University Press.

Lenoir, T. (1982) *The Strategy of Life: Teleology and Mechanics in Nineteenth-Century German Biology*, Chicago, University of Chicago Press.

Lindeman, R. L. (1942) "The trophic dynamic aspect of ecology," *Ecology*, 23, pp. 399–419.

Loison, L. (2010) *Qu'est-ce que le néolamarckisme? Les biologistes français et la question de l'évolution des espèces*, Paris, Vuibert.

Loison, L. (2012) "Le concept de cellule chez Claude Bernard et la constitution du transformisme expérimental," in J.-G. Barbara and P. Corvol (eds.), *Les élèves de Claude Bernard: Les nouvelles disciplines physiologiques en France au tournant du XXe siècle*, Paris, Hermann, pp. 135–149.

Lombard, J. (2014) "Once upon a time the cell membranes: 175 years of cell boundary research," *Biol. Direct*, 9, art. no. 32.

Lopez-Beltran, C. (2004) "In the cradle of heredity: French physicians and *l'hérédité naturelle* in the early nineteenth century," *J. Hist. Biol.*, 37, pp. 39–72.

Lotka, J. A. (1922a) "Contribution to the energetics of evolution," *Proc. Natl. Acad. Sci. USA*, 8, pp. 147–151.

Lotka, J. A. (1922b) "Natural selection as a physical principle," *Proc. Natl. Acad. Sci. USA*, 8, pp. 151–154.

Lovelock, J. (2000) *Homage to Gaia: The Life of an Independent Scientist*, Oxford, Oxford University Press.

Lucretius (1995), *On the Nature of Things*, Baltimore, MD, Johns Hopkins University Press.

Lwoff, A. (1944) *L'évolution physiologique: Études des pertes de fonctions chez les microorganismes*, Paris, Hermann.

Lwoff, A. (1981) *Jeux et combats*, Paris, Fayard.

Lynch, M. (2007) *The Origins of Genome Architecture*, Sunderland, MA, Sinauer.

Lyons, S. (1999) *Thomas Henry Huxley: The Evolution of a Scientist*, Amherst, NY, Prometheus Books.

MacArthur, R. H., and Wilson, E. O. (1967) *The Theory of Island Biogeography*, Princeton, NJ, Princeton University Press.

Magner, L. N. (2002) *A History of the Life Sciences*, New York, Marcel Dekker.

Maienschein, J. (1978) "Cell lineage, ancestral reminiscence, and the biogenetic law," *J. Hist. Biol.*, 11, pp. 129–158.

Maienschein, J. (2014) *Embryos under the Microscope: The Diverging Meanings of Life*, Cambridge, MA, Harvard University Press.

Mandressi, R. (2003) *Le regard de l'anatomiste: Dissections et invention du corps en Occident*, Paris, Éditions du Seuil.

Maruyama, K. (1991) "The discovery of adenosine triphosphate and the establishment of its structure," *J. Hist. Biol.*, 24, pp. 145–154.

Maupertuis (1980) *Vénus physique suivie de la Lettre sur le progrès des sciences* [1745], Paris, Aubier-Montaigne.

Maupertuis (2001) *Système de la nature*, in *Œuvres de M. de Maupertuis*, vol. II [1754], Boston, MA, Adamant Media.

Maynard Smith, J. (1982) *Evolution and the Theory of Games*, Cambridge, Cambridge University Press.

Mayr, E. (1961) "Cause and effect in biology," *Science*, 134, pp. 1501–1506.

Mayr, E. (1982) *The Growth of Biological Thought: Diversity, Evolution, and Inheritance*, Cambridge, MA, Harvard University Press.

McIntosh, R. (1985) *The Background of Ecology*, Cambridge, Cambridge University Press.

McLaughlin, P. (2002) "Naming biology," *J. Hist. Biol.*, 35, pp. 1–4.

Medawar, P. B. (1952) *An Unsolved Problem of Biology*, London, H. K. Lewis for University College London.

Méthot, P. O. (2016) "Bacterial transformation, and the origins of epidemics in the interwar period: The epistemological significance of Fred Griffith's transforming experiment," *J. Hist. Biol.*, 49, pp. 311–358.

Meulders, M. (2010) *Helmholtz: From Enlightenment to Neuroscience*, Cambridge, MA, MIT Press.

Michelet, J. (1858) *L'insecte*, Paris, Librairie de L. Hachette.

Millstein, R. L. (2008) "Distinguishing drift and selection empirically: 'The great snail debate' of the 1950s," *J. Hist. Biol.*, 41, pp. 339–367.

Mitchell, S. (2002) "Integrative pluralism," *Biol. Philos.*, 17, pp. 55–70.

Monod, J., Wyman, J. and Changeux, J.-P. (1965) "On the nature of allosteric transition: A plausible model," *J. Mol. Biol.*, 12, pp. 88–118.

Montgomery, W. M. (1970) "The origins of the spiral theory of phyllotaxis," *J. Hist. Biol.*, 3, pp. 299–323.

Morange, M. (1993) "The discovery of cellular oncogenes," *Hist. Philos. Life Sci.*, 15, pp. 45–58.

Morange, M. (2005) *Les secrets du vivant: Contre la pensée unique en biologie*, Paris, La Découverte.

Morange, M. (2006a) "Ciliates as models . . . of what?" *J. Biosci.*, 31, pp. 27–30.

Morange, M. (2006b) "Émile Duclaux (1840–1904)," *J. Biosci.*, 31, pp. 215–218.

Morange, M. (2007) "z-DNA: The nature is not opportunistic," *J. Biosci.*, 32, pp. 657–661.

Morange, M. (2008a) "Regulation of gene expression by non-coding RNAs: The early steps," *J. Biosci.*, 33, pp. 327–331.

Morange, M. (2008b) *À quoi sert l'histoire des sciences?* Paris, Quae.

Morange, M. (2011a) "Construction of the ribbon model of proteins (1981): The contribution of Jane Richardson," *J. Biosci.*, 30, pp. 571–574.

Morange, M. (2011b) "From Mechnikov to proteotoxicity: Ageing as the result of an intoxication," *J. Biosci.*, 36, pp. 769–772.

Morange, M. (2012) "Transfers from plant biology: From cross protection to RNA interference and DNA vaccination," *J. Biosci.*, 37, pp. 949–952.

Morange, M. (2020) *The Black Box of Biology: A History of the Molecular Revolution*, trans. M. Cobb, Cambridge, MA, Harvard University Press.

Morgan, G. J. (2013) "Linus Pauling: Leading exporter of chemical insights into biology," in O. Harman and M. R. Dietrich (eds.), *Outsider Scientists: Routes to Innovation in Biology*, Chicago, University of Chicago Press, pp. 110–127.

Morris, P. J. (1997) "Louis Agassiz's arguments against Darwinism in his additions to the French translation of the Essay on classification," *J. Hist. Biol.*, 30, pp. 121–134.

Moulin, A.-M. (1991) *Le dernier langage de la médecine: Histoire de l'immunologie, de Pasteur au SIDA*, Paris, Presses universitaires de France.

Müller-Wille, S., and Charmantier, I. (2012) "Natural history and information overload: The case of Linnaeus," *Stud. Hist. Philos. Biol. Biomed. Sci.*, 43, pp. 4–15.

Müller-Wille, S., and Rheinberger, H.-J. (eds.) (2007) *Heredity Produced*, Cambridge, MA, MIT Press.

Müller-Wille, S., and Rheinberger, H.-J. (2012) *A Cultural History of Heredity*, Chicago, University of Chicago Press.

Nagy, Z. (2014) *A History of Modern Immunology: The Path Toward Understanding*, London, Academic Press.

Nelson, G. (1978) "From Candolle to Croizat: Comments on the history of biogeography," *J. Hist. Biol.*, 11, pp. 269–305.

Nordenskiöld, E. (1928), *The History of Biology: A Survey*, New York, Alfred A. Knopf.

Normandin, S., and Wolfe, C. T. (2013) *Vitalism and the Scientific Image in Post-Enlightenment Life Science, 1800–2010*, New York, Springer.

Nouvel, P. (ed.) (2011) *Repenser le vitalisme*, Paris, Presses universitaires de France.

Nyhart, L. K. (1995) *Biology Takes Form: Animal Morphology and the German Universities 1800–1900*, Chicago, University of Chicago Press.

Odenbaugh, J. (2013) "Searching for patterns, hunting for causes: Robert MacArthur, the mathematical naturalist," in O. Harman and M. R. Dietrich (eds.), *Outsider Scientists: Routes to Innovation in Biology*, Chicago, University of Chicago Press, pp. 181–198.

Olby, R. (1974) *The Path to the Double Helix*, London, Macmillan.

Olby, R. (1979) "Mendel no mendelian?" *Hist. Sci.*, 17, pp. 53–72.

Olby, R. (2009) *Francis Crick, Hunter of Life's Secrets*, Cold Spring Harbor, NY, Cold Spring Harbor Laboratory.

Orbigny, A. Dessalines d' (1860) *Paléontologie française* [1842], Paris, Arthus Bertrand.

Oshinsky, D. M. (2005) *Polio: An American Story*, New York, Oxford University Press.

Otis, L. (2007) *Müller's Lab*, Oxford, Oxford University Press.

Pagel, W. (1982) *Joan Baptista van Helmont: Reformer of Science and Medicine*, Cambridge, Cambridge University Press.

Paracelsus (2008) *Essential Theoretical Writings*, ed. and trans. Andrew Weeks, Leiden, Brill.

Pasteur, L. (1993) *Écrits scientifiques et médicaux*, Paris, Flammarion.

Pauly, P. J. (1987) *Controlling Life: Jacques Loeb and the Engineering Ideal in Biology*, Oxford, Oxford University Press.

Perrier, E. (1888) *Le transformisme*, Paris, Baillière et fils.

Peterson, E. L. (2008) "William Bateson from *Balanoglossus* to *Materials for the Study of Variation*: The transatlantic roots of discontinuity and the (un)naturalness of selection," *J. Hist. Biol.*, 41, pp. 267–305.

Pichot, A. (1993) *Histoire de la notion de vie*, Paris, Gallimard.

Pichot, A. (1999) *Histoire de la notion de gène*, Paris, Flammarion.

Pichot, A. (2011) *Expliquer la vie: De l'âme à la molécule*, Versailles, Quae.

Pinon, L. (1995) *Livres de zoologie de la Renaissance: Une anthologie*, Paris, Klincksieck.

Piveteau, J. (1950) "Le débat entre Cuvier et Geoffroy Saint-Hilaire sur l'unité de plan et de composition," *Rev. Hist. Sci.*, 3, pp. 343–363.

Podolsky, S. (1996) "The role of the virus in origin-of-life theorizing," *J. Hist. Biol.*, 29, pp. 79–126.

Provine, W. B. (1971) *The Origins of Theoretical Population Genetics*, Chicago, University of Chicago Press.

Rashed, R. (ed.) (1997) *Technologie, alchimie, et sciences de la vie*, vol. 3 of *Histoire des sciences arabes*, Paris, Éditions du Seuil.

Rasmussen, N. (1999) "The forgotten promise of thiamin: Merck, Caltech biologists, and plant hormones in a 1930s biotechnology project," *J. Hist. Biol.*, 32, pp. 245–261.

Rasmussen, N. (2014) *Gene Jockeys: Life Science and the Rise of Biotech Enterprise*, Baltimore, MD, Johns Hopkins University Press.

Rather, L. J. (1978) *The Genesis of Cancer: A Study in the History of Ideas*, Baltimore, MD, Johns Hopkins University Press.

Reill, P. H. (2005) *Vitalizing Nature in the Enlightenment*, Berkeley, University of California Press.

Reisse, J. (2013) *Wallace: Alfred Russel Wallace, plus darwiniste que Darwin mais politiquement moins correct*, Bruxelles, Académie royale de Belgique.

Rheinberger, H. J. (2000a) "Mendelian inheritance in Germany between 1900 and 1910: The case of Carl Correns," *C. R. Acad. Sci. Paris Sci. Vie*, 323, pp. 1089–1096.

Rheinberger, H.-J. (2000b) "Ephestia: Le projet expérimental d'une génétique de la physiologie du développement dans l'œuvre d'Alfred Kühn," *Rev. Hist. Sci.*, 53, pp. 401–446.

Richards, R. J. (2002) *The Romantic Conception of Life: Science and Philosophy in the Age of Goethe*, Chicago, University of Chicago Press.

Richmond, M. L. (2000) "T. H. Huxley's criticism of German cell theory: An epigenetic and physiological interpretation of cell structure," *J. Hist. Biol.*, 33, pp. 247–289.

Richmond, M. L. (2007) "Muriel Wheldale Onslow and early biochemical genetics," *J. Hist. Biol.*, 40, pp. 389–426.

Riddle, J. M. (2013) *Dioscorides on Pharmacy and Medicine*, Austin, University of Texas Press.

Roe, S. A. (1981) *Matter, Life and Generation: Eighteenth Century Embryology and the Haller-Wolff Debate*, Cambridge, Cambridge University Press.

Roger, J. (1963) *Les sciences de la vie dans la pensée française du XVIIIe siècle*, Paris, Armand Colin.

Roger, J. (1989) *Buffon*, Paris, Fayard.

Roll-Hansen, N. (1979), "Experimental method and spontaneous generation: The controversy between Pasteur and Pouchet, 1859–1864," *J. Hist. Med.*, 34, pp. 273–292.

Roll-Hansen, N. (2004) *The Lysenko Effect: The Politics of Science*, New York, Humanity Books.

Rostand, J. (1945) *Esquisse d'une histoire de la biologie*, Paris, Gallimard.

Rostand, J. (1951) *Les origines de la biologie expérimentale et l'Abbé Spallanzani*, Paris, Fasquelle.

Rudwick, M.J.S. (2005) *Bursting the Limits of Time: The Reconstruction of Geohistory in the Age of Revolution*, Chicago, University of Chicago Press.

Rudwick, M.J.S. (2014) *Earth's Deep History: How It Was Discovered and Why It Matters*, Chicago, University of Chicago Press.
Salomon-Bayet, C. (1978) *L'institution de la science et l'expérience du vivant: Méthode et expérience à l'Académie royale des sciences, 1666–1793*, Paris, Flammarion.
Sandler, I. (1983) "Pierre Louis Moreau de Maupertuis: A precursor of Mendel?" *J. Hist. Biol.*, 16, pp. 101–136.
Sapp, J. (1987) *Beyond the Gene: Cytoplasmic Inheritance and the Struggle for Authority in Genetics*, Oxford, Oxford University Press.
Sapp, J. (1994) *Evolution by Association: A History of Symbiosis*, Oxford, Oxford University Press.
Sapp, J. (2003) *The Evolution of Biology*, Oxford, Oxford University Press.
Schmitt, S. (2006) *Aux origines de la biologie moderne: L'anatomie comparée d'Aristote à la théorie de l'évolution*, Paris, Belin.
Schmitt, S. (ed.) (2007) *Buffon: Œuvres*, Paris, Gallimard, "La Pléiade."
Schmitt, S. (ed.) (2013) *Pline l'Ancien: Histoire naturelle*, Paris, Gallimard, "La Pléiade."
Secord, J. A. (2000) *Victorian Sensation: The Extraordinary Publication, Reception, and Secret Authorship of Vestiges of the Natural History of Creation*, Chicago, University of Chicago Press.
Sepkoski, D. (2008) "Stephen Jay Gould, Darwinian iconoclast ?" in O. Harman and M. R. Dietrich (eds.), *Rebels, Mavericks, and Heretics in Biology*, New Haven, CT, Yale University Press, pp. 321–337.
Sepkoski, D. (2012) *Rereading the Fossil Record. The Growth of Paleobiology as an Evolutionary Discipline*, Chicago, University of Chicago Press.
Serafini, A. (2001) *The Epic History of Biology*, New York, Basic Books.
Sherrington, C. (1906) *The Integrative Action of the Nervous System*, New Haven, CT, Yale University Press.
Shubin, N., Tabin, C., and Carroll, S. (2009) "Deep homology and the origins of evolutionary novelty," *Nature*, 457, pp. 818–823.
Silverstein, A. M. (2003) "Cellular versus humoral immunology: A century-long dispute," *Nat. Immunol.*, 4, pp. 425–428.
Simpson, G. G. (1944) *Tempo and Mode in Evolution*, New York, Columbia University Press.
Singleton, R., Jr., and Singleton, D. R. (2017) "Remembering our forebears: Albert Jan Kluyver and the unity of life," *J. Hist. Biol.*, 50, pp. 169–218.
Slack, N. G. (2003) "Are research schools necessary? Contrasting models of twentieth century research at Yale led by Ross Granville Harrison, Grace E. Pickford and G. Evelyn Hutchinson," *J. Hist. Biol.*, 36, pp. 501–529.
Slack, N. G. (2011) *G. Evelyn Hutchinson and the Invention of Modern Ecology*, New Haven, CT, Yale University Press.

Sloane, P. R. (1972) "John Locke, John Ray, and the problem of the natural system," *J. Hist. Biol.*, 5, pp. 1–53.

Slobodchikoff, C. N. (1976) *Concepts of Species*, Stroudsburg, PA, Dowden, Hutchinson and Ross.

Snell, G. D., and Reed, S. (1993) "William Ernest Castle, pioneer mammalian geneticist," *Genetics*, 133, pp. 751–753.

Stamhuis, I. H., Meijer, O. G., and Zevenhuisen, E.J.A. (1999) "Hugo de Vries on heredity, 1889–1903: Statistics, Mendelian laws, pangenes, mutations," *Isis*, 90, pp. 238–267.

Stevens, H. (2013) *Life Out of Sequence: A Data-driven History of Bioinformatics*, Chicago, University of Chicago Press.

Strick, J. E. (2000) *Sparks of Life: Darwinism and the Victorian Debate over Spontaneous Generation*, Cambridge, MA, Harvard University Press.

Sturtevant, A. H. (1965) *A History of Genetics*, New York, Harper and Row.

Suarez, E., and Barahona, A. (1996) "The experimental roots of the neutral theory of molecular evolution," *Hist. Philos. Life Sci.*, 18, pp. 55–81.

Summers, W. C. (1999) *Félix d'Herelle and the Origins of Molecular Biology*, New Haven, CT, Yale University Press.

Sunderland, M. E. (2010) "Regeneration: Thomas Hunt Morgan's window into development," *J. Hist. Biol.*, 43, pp. 325–361.

Swazey, J. P. (1968) "Sherrington's concept of integrative action," *J. Hist. Biol.*, 1, pp. 57–89.

Tanford, C., and Reynolds, J. (2001) *Nature's Robots: A History of Proteins*, Oxford, Oxford University Press.

Taquet, P. (2006) *Georges Cuvier: Naissance d'un génie*, Paris, Odile Jacob.

Tauber, A. I. (2003) "Metchnikoff and the phagocytosis theory," *Nat. Rev. Mol. Cell Biol.*, 4, pp. 897–901.

Theophrastus (1976–1990) *De causis plantarum*, trans. B. Einarson and G.K.K. Link, Cambridge, MA, Harvard University Press.

Theophrastus (2014) *Enquiry into Plants*, trans. A. F. Hort, Cambridge, MA, Harvard University Press.

Thompson, D'A. (2014) *On Growth and Form*, Cambridge, Cambridge University Press.

Thurtle, P. (2007) *The Emergence of Genetic Rationality*, Seattle, University of Washington Press.

Tirard, S. (2010) *Histoire de la vie latente: Des animaux ressuscitants du XVIIIe siècle aux embryons congelés du XXe siècle*, Paris, Vuibert.

Trumpler, M. (1997) "Converging images: Techniques of intervention and forms of representation of sodium-channel proteins in nerve cell membranes," *J. Hist. Biol.*, 30, pp. 55–89.

Tucker, H. (2011) *Blood Work: A Tale of Medicine and Murder in the Scientific Revolution*, New York, W. W. Norton.

Turing, A. (1952) "The chemical basis of morphogenesis," *Philos. Trans. R. Soc. Lond. B*, 237, pp. 37–72.

Vallade, J. (2008) *L'œil de lynx des microscopistes*, Dijon, Éditions universitaires de Dijon.

Van der Lugt, M., and de Miramon, C. (eds.) (2008) *L'hérédité entre Moyen Âge et époque moderne: Perspectives historiques*, Florence, SISMEL Edizioni del Galluzzo.

Van der Valk, A. G. (2014) "From formation to ecosystems: Tansley's response to Clements' climax," *J. Hist. Biol.*, 47, pp. 293–321.

Waddington, C. H. (1942) "L'épigénotype," *Endeavour*, 1, pp. 18–20.

Williams, G. C. (1966) *Adaptation and Natural Selection*, Princeton, NJ, Princeton University Press.

Wilson, E. O. (2000) *Sociobiology: The New Synthesis* [1975], Cambridge, MA, Harvard University Press.

Witkowski, J. A. (1986) "Ross Harrison and the experimental analysis of nerve growth: The origins in tissue culture," in T. J. Horder, J. A. Witkowski, and C. C. Wylie (eds.), *A History of Embryology*, Cambridge, Cambridge University Press, pp. 149–177.

Wolfe, E. L., Barger, A. C., and Benson, S. (2000) *Walter B. Cannon: Science and Society*, Cambridge, MA, Harvard University Press.

Wolf-Ernst, R. (1986) "The search for a macroevolutionary theory in German paleontology," *J. Hist. Biol.*, 19, pp. 79–130.

Wolff, C. F. (2003) *De formatione intestinorum: Sur la formation des intestins* [1768–1769], Brussels, Brepols.

Wood, R., and Orel, V. (2001) *Genetic Prehistory in Selective Breeding: A Prelude to Mendel*, Oxford, Oxford University Press.

Wynne-Edwards, V. C. (1962) *Animal Dispersion in Relation to Social Behaviour*, Edinburgh, Oliver and Boyd.

Zuckerkandl, E., and Pauling, L. (1965) "Evolutionary divergence and convergence in proteins," in V. Bryson and H. J. Vogel (eds.) *Evolving Genes and Proteins*, New York, Academic Press, pp. 97–166.

INDEX OF NAMES

THEMATIC INDEX

A NOTE ON THE TYPE

This book has been composed in Arno, an Old-style serif typeface in the classic Venetian tradition, designed by Robert Slimbach at Adobe.